P9-EKC-420

THE GOOD GUT

JUSTIN SONNENBURG, PhD,

and ERICA SONNENBURG, PhD

PENGUIN PRESS

NEW YORK

2015

THE GOOD GUT

Taking Control of Your Weight, Your Mood, and Your Long-term Health

PENGUIN PRESS
Published by the Penguin Publishing Group
Penguin Random House LLC
375 Hudson Street
New York, New York 10014

USA · Canada · UK · Ireland · Australia
New Zealand · India · South Africa · China

penguin.com
A Penguin Random House Company

First published by Penguin Press, an imprint of Penguin Publishing Group,
a division of Penguin Random House LLC, 2015

Photograph credits
Pages 12 and 14: © Justin Sonnenburg, Jaime Dant, and Jeffrey Gordon
Page 17: Pascal Gagneux
Page 74: © Kristen Earle and Justin Sonnenburg

ISBN 978-1-59420-628-3

Printed in the United States of America
10 9 8 7 6 5 4 3 2 1

DESIGNED BY MARYSARAH QUINN

TO OUR DAUGHTERS, CLAIRE AND CAMILLE,
who provided the inspiration to write this book,
and
to our trillions of microbial inhabitants—may your secrets continue to enlighten us for years to come.

CONTENTS

FOREWORD • IX

INTRODUCTION • 1

1. What Is the Microbiota and Why Should I Care? • 9
2. Assembling Our Lifelong Community of Companions • 35
3. Setting the Dial on the Immune System • 61
4. The Transients • 85
5. Trillions of Mouths to Feed • 111
6. A Gut Feeling • 137
7. Eat Sh!t and Live • 163

8. The Aging Microbiota • 187

9. Managing Your Internal Fermentation • 209

MENUS AND RECIPES • 227

ACKNOWLEDGMENTS • 267

APPENDIX • 269

NOTES • 271

BIBLIOGRAPHY • 283

INDEX • 293

FOREWORD

ANDREW WEIL, MD

In medical school in the mid-1960s I learned that the human colon contained a great many bacteria necessary for proper digestion and the assimilation of nutrients and that prolonged use of antibiotics could cause intestinal disturbances as a result of the overgrowth of undesirable organisms. In that era, people who ate yogurt for digestive health or took acidophilus supplements were labeled "health nuts," and few medical authorities believed that gut flora had any influence outside of the gastrointestinal tract. There was no concept of a human microbiome that comprised all the microorganisms in and on the body, whose total DNA content was greater than that of human DNA.

Today, research on the human microbiome is one of the hottest areas of medical science, heralding a true revolution in our understanding of physiology and offering great promise for optimizing health and managing disease in novel ways. The species of bacteria and fungi that colonize the gut may determine our interactions with the environment, protecting us from or predisposing us to the development of allergy and autoimmunity. They may protect us from or predispose us to becoming obese or diabetic. They may inhibit or intensify inflammation in the body. They may interact with artificial sweeteners to cause insulin resistance and weight gain in some individuals. They may even influence mental function and emotional wellness.

I first heard about this new view of the microbiome from one of the authors of this book, Justin Sonnenburg. He and his wife, Erica, are prominent researchers in the field, directing their own lab in the Department of Microbiology and Immunology at Stanford University School of Medicine. In 2013, I had invited Justin to give a plenary talk on his research at the 10th Annual Nutrition and Health Conference, organized by the University of Arizona Center for Integrative Medicine. The conference took place in Seattle and was attended by hundreds of physicians, registered dieticians, and other health professionals. Justin's talk was the highlight of the event for me. It conveyed the excitement of discoveries about the human microbiome and suggested answers to puzzling questions I had about health conditions that are on the rise.

Asthma, allergies, and autoimmunity have all increased in North America and other developed areas of the world. Why is the incidence of peanut allergy so much greater today than it was when I grew up in the 1950s? And what is the explanation for the spectacular increase in gluten sensitivity?

That last question had bothered me greatly. Granted that gluten intolerance is a patient-driven diagnosis for which objective tests are lacking, there are still more and more people whose symptoms resolve when gluten is removed from their diet and reappear when it is added back. I reject the notion that grains in general and wheat in particular are bad foods, and I am unconvinced by the argument that the genetics of wheat have changed enough in recent years to be the cause. Gluten sensitivity seems to be most prominent in the North

American population. In China, where isolated gluten is served at most restaurants—in such dishes as gluten with black bean sauce, and sweet-and-sour gluten—sensitivity to this protein is unknown, as it is in Japan. What has changed in North America that might explain it?

Justin Sonnenburg gave me the insight that alterations in our microbiome are likely to blame. Four factors have greatly changed gut flora in individuals in our population over the past few decades. They are: 1) increasing consumption of industrialized, processed foods, 2) widespread use of antibiotics, 3) the alarming rise in Caesarean deliveries, now accounting for one in every three births, and 4) the decline in breast-feeding. In this book you will learn how each of these has contributed to drastic alterations in the human microbiome and how those alterations may account for the increasing incidence of diverse chronic health conditions including autism, depression, and other mental/emotional disturbances.

The Sonnenburgs also discuss possibilities for typing the microbiome as a new diagnostic modality, and they examine the very important question of whether (and how) we might modify our microbiome to reduce disease risks and optimize health. How to do so is a very individual question, and the answer changes as you age. Should you take probiotic dietary supplements? Do they work? Which ones are most effective? What about fermented foods, so prominent in the diets of East Asians? (I believe we should be making and eating more of them.) This book tells you how to tackle all of these questions.

I consider this work essential reading for all health professionals and for everyone interested in a broader understanding of health and wellness. I am sure that you will come away from it with the same excitement that the authors and I feel over these new discoveries about the microorganisms that are so much a part of us.

Tucson, Arizona
October 2014

THE GOOD GUT

INTRODUCTION

We all know that much of our health is predetermined by our genes. We also know that we can generally improve our health if we eat right, exercise, and manage our stress. But how to do those things is a matter of great debate. Many well-meaning health programs are focused solely on weight loss or heart health, but what if there was another key to our overall health, a second, malleable genome that could influence our weight, our mood, and our long-term well-being? What if we could influence this genome by very specific (and often surprising) lifestyle choices? Well, this second genome exists. It belongs to the bacteria that inhabit our gut and is vital in countless ways to our overall well-being. The details of how this microbial community, known as the microbiota, is hard-wired into health and disease are starting to come to light and they are reshaping what it means to be human.

As scientists try to unravel the causes behind the prevalence of predominantly Western afflictions such as cancer, diabetes, allergies, asthma, autism, and inflammatory bowel diseases, it is becoming clear that the microbiota plays an important role in the development of each of these conditions and in many other aspects of our health. Our bacterial inhabitants touch all aspects of our biology in some way, directly or indirectly.

The inhabitants of our gut have evolved inside us over millennia, but today they face new challenges. The modern world has changed the way we eat (highly processed, calorie-dense, industrially produced foods) and how we live (homes sanitized with antibacterial cleaners and the overuse of antibiotics), and these changes threaten the health of our intestinal microbiota.

Our digestive system is much more than a collection of cells that surround our last few meals—it also contains a dense consortium of bacteria and other microbes. Although all body surfaces, orifices, and cavities are teeming with microbes, the vast majority are located within our large intestine. Among their many attributes, these bacteria chemically snip and consume indigestible dietary fiber and convert it into compounds that our colon absorbs. Some of these compounds are vital to our health and are our last chance at salvaging nutrients from hard-to-digest dietary fiber. Nurturing our gut bacteria so that they produce the compounds that our bodies need is one of the most important choices we can make for our health.

More than we ever suspected, our gut microbiota sets the dial on our immune system. Our immune system is central to all aspects of our health. When it works well, we fight off infections efficiently and extinguish malignancies at their earliest appearance. When the immune system operates suboptimally, numerous ailments can result. If the gut bacteria are healthy, it's likely that the immune system is running well. If the gut bacteria are not healthy, we are at increased risk of developing autoimmune disease and cancer. Microbiota-produced chemicals can impact the level of inflammation—our immune system's response to injury or perceived threats that manifests as swelling, redness, and irritation—in our gut and throughout our body.

Inflammation can have a cascading effect on all manner of health issues.

Some of the chemicals produced by the microbiota even communicate directly with our central nervous system through the brain-gut axis. We are still learning much about how the microbiota impacts our brains. The brain-gut axis influences our well-being profoundly, doing far more than just letting us know when it's time to eat. Gut bacteria can affect moods and behavior and may alter the progression of some neurological conditions.

Every person's alliance with microbes begins at birth. Although we are sterile within the womb, upon arriving in the outside world microbes rapidly colonize the body's virgin habitat. These microbes come from our mothers, friends and family members, and the environment. As noted biologist Stan Falkow once said, "The world is covered in a patina of shit." Or, if you like, it's covered in a patina of bacteria. That's not a bad thing. So the next time your infant sticks a new object into his or her mouth, if it is not a choke hazard, instead of rushing to pull it out or clean it with a sanitizer, consider how the bacterial patina is providing valuable microbes to help form the new microbiota. As life progresses, our resident microbial communities are shaped by factors such as whether we were born naturally or by C-section, if we were breast- or formula fed, how often we use antibiotics, if we own a dog, and by the food we eat.

The mounting evidence that these bacteria are paramount to our health and well-being means that the lifestyle, medical, and dietary choices we make need to include careful consideration of the consequences to our gut microbes. Twenty-first-century DNA sequencing technology offers a detailed view of our microbiota's more than 2 million microbial genes—called the microbiome—and a few striking themes have become apparent. First, each of us has a microbiota as unique as our fingerprint that impacts our predisposition to different diseases. Second, the microbiota can malfunction and contribute to the development of diseases and conditions, such as obesity, that we once thought were attributable solely to lifestyle. And third, because of the microbiota's capacity for change it enables us to alter our overall health as we age.

Proper care of and appreciation for the microbiota is essential for good health. We can use this new information to answer many questions, including: How can we guide microbiota assembly at birth so children get on the path to having a healthy microbiota? How can we optimize our microbiota through adulthood to strengthen our immune system and decrease the risk of autoimmune diseases and allergies? What specific dietary changes can we make to nurture our microbiota? When we need to take a course of antibiotics, how can we regain a flourishing microbiota? How can we minimize the decline that occurs in the microbiota as we age? How can we find the right combination of microbes for our own personal gut?

While there is still a lot to uncover about the microbiota, the past decade has seen an explosion in our understanding of this community of microbes and how they link to human health and disease. Ten years ago, it was clear that the microbiota represented an important, albeit poorly understood, feature of human biology. The richness of unanswered questions offered fertile territory for beginning a career as a biomedical scientist; and it was clear that this subject was going to be pivotal to many aspects of human health.

Our gut is home to more than 100 trillion bacteria. If you lined up all of your bacteria end to end, they would reach the moon. These bacteria are found throughout our digestive system and, depending on their type, may decide to live in your stomach (although most don't because of the stomach's harsh and acidic climate) or small intestine, but most end up residing in the large intestine. Hundreds of species of bacteria, together totaling in the trillions, live in the large intestine at a density of 500 billion cells per teaspoon of intestinal contents.

Clearly there is no shortage of bacteria in our gut, which can make this next statement a little hard to believe. Our gut bacteria belong on the endangered species list. The average American adult has approximately 1,200 different species of bacteria residing in his or her gut. That may seem like a lot until you consider that the average Amerindian living in the Amazonas of Venezuela has roughly 1,600 species, a full third more. Similarly, other groups of humans with lifestyles and diets more similar to our ancient human ancestors have more varied bacteria in their gut than we Americans do. Why is this happening? Our overly processed Western diet, overuse of antibiotics, and sterilized homes are threatening the health and stability of our intestinal inhabitants.

If your gut bacteria were able to walk through your average grocery store tasked with finding something to eat, they would face the equivalent of humans trying to find food in a Home Depot. The candy racks by the register don't count because they, as Michael Pollan so perfectly stated, are filled not with food but "food-like substances." Thanks to our typical diet, the average American's gut bacteria are starving. To add insult to injury, a couple of times a year we are prescribed gut bacteria poison, or antibiotics, as they are commonly called. To top it off we spend an average of almost $700 a year on household cleaners so that our homes are nearly as sterile as an operating room. And don't forget the ubiquitous bottles of hand sanitizer found at grocery store entrances, on library counters, and even hanging from school backpacks.

It's hard to know for sure where this path is leading us. In the near future, might we have half of the bacterial species our ancestors had, or even less? If so, what will this mean for us? We have already begun to see the effects of the Western lifestyle on our health in terms of obesity, diabetes, and autoimmune diseases. These diseases are not typically found in societies with a more diverse microbiota. Will these diseases become even more prevalent, appear earlier in life, or spread across the globe as the world adopts our microbiota-attacking lifestyle? It is possible that gut bacterial species that make important contributions to our health will become extinct, or so rare that our microbiota won't resemble that of early humans, and maybe this has already happened to some extent.

We have become a nation of junk-food addicts and we are indoctrinating our youth into this extremely hazardous existence—they are the unwitting victims of our microbiota-harming lifestyle and it is making them sick and will shorten their life span.

As scientists we write papers about our research on the microbiota, but this information is conveyed in a way that is not easily accessible to the public at large. In other words, it's geeky. Scientists are trained to be highly skeptical, so it is not in our nature to provide recommendations unless they have been put through the rigors of a double-blind, placebo-controlled study. But within our own family, we had already made changes in our diet and lifestyle based on the findings in our lab and the labs of others studying the microbiota. As our daughters grew and we interacted with other families with young kids, we saw parents trying to make informed decisions about food. However, they weren't taking into account a central element of health, development of their child's microbiota. How could they if the information was not available to them? We were aware that we had very unique insight into the biology of the digestive tract and its resident microbes, insight that was clearly guiding many of our decisions related to how we feed ourselves and our children, in addition to guiding other aspects of our lifestyle.

We committed ourselves to writing this book with the hope of compiling the essential information nonscientists need in order to make sense of the flurry of new microbiota research findings. We have used the current available data to provide practical advice and suggestions for dietary and lifestyle choices—a way to optimize health with a focus on the entity on which so much of our biology is centered—the gut microbiota.

We have organized the book to lead you through the most interesting and relevant findings within our field to show how they impact you throughout life. We will look at what the microbiota is and how it colonizes us; how we can nourish it; what its amazing properties are; what the great frontiers of this field are; how the microbiota ages; and how to care for it throughout life.

After a brief orientation to the microbiota we explain the development of the gut microbiota starting with our sterile digestive tract just before birth and progressing through infancy and childhood. This section includes suggestions for how to ensure children adopt microbiota-supporting eating habits as they make the transition to solid food and is a must-read for new or expectant parents who are thinking about how to get their child set on a trajectory for long-term health. Later chapters delve into the connections that have been made between our microbiota and our immune system and metabolism. We address how modern societies are making many mistakes in caring for the gut microbiota, and discuss ways to reform our diet and lifestyle to help our microbiota and thus promote our health and combat the onset of chronic diseases. We touch on the exciting emerging connection between the gut microbiota and the brain, including the latest findings in this fast-moving arena of research that links the microbiota to mood and behavior. In Chapter 7 we describe the latest breakthroughs in the treatment of problematic microbiotas to restore health (which includes reprogramming diseased microbiotas using fecal transplant) and discuss the bright future in this new area of therapeutic discovery. The recently documented decline of the microbiota that occurs as we age is the focus of Chapter 8, along with suggestions for how this decline can be minimized to improve digestive health and overall well-being in seniors. Finally, we bring together all of the practical advice found throughout the book into one discrete plan for getting your microbiota on the right track and maintaining a state that maximizes health benefits in the long term. This final chapter includes recipes and meal plans to aid the busiest of people and families to attain microbiota-aided health efficiently and deliciously.

We must emphasize that the field of microbiota research is still in its infancy or, at best, its toddlerhood, but we can certainly use our present understanding of the microbiota to guide decisions within our own lives; and we feel that sufficient information exists to make general recommendations. It is important that individuals consult their physicians before implementing these recommendations, particularly if special health considerations exist.

Our goal is also to educate you on the critical role that this community plays in your overall health. We hope this book will give readers a platform for interpreting and understanding new revelations and enable them to incorporate these insights into their choices about their diet and general lifestyle. Unlike the human genome, which is largely fixed before birth, the microbiome can be altered throughout life by way of strategic choices that are within our control. The microbiome's plasticity provides us with a huge opportunity to shape it in a way that optimizes our health.

As composite organisms containing both human and microbial parts, we must recognize that their biology is intimately intertwined. These microbes are our partners throughout life, and if we can nurture and care for them, they will in turn protect us, the human bodies that they call home.

CHAPTER 1

What Is the Microbiota and Why Should I Care?

THE MICROBIAL WORLD

We like to think of the world as being dominated by humans. Our species has created complex societies, built elaborate cities, and produced amazing works of art, music, and literature. Evidence of human activity on this planet, such as highways, dams, and illuminated skylines, is even visible from space! While it is clear we've had a large impact on Earth, the reality is that humans are relatively new and numerically minor inhabitants of our planet. We live in a microbial world. The earth is covered in microorganisms, or microbes, and has been for billions of years. Microbes are microscopic life such as bacteria and archaea. There are more microbes present on your hand than there are people in the world. If you lumped all the bacteria on Earth together they would form a biomass larger than all the plants and animals combined. (Keep this mental image in mind for an update on our antibiotic war with these microbes as described in the pages to come.) One estimate places the number of bacteria on Earth at 5 million trillion trillion or, in geekier terms, 5 nonillion. If you want to write it out, it's a 5 with 30 zeros after it.

Bacteria are everywhere, from cold, dark lakes buried a half mile under the Antarctic ice to deep-sea hydrothermal vents reaching temperatures over 200°F to the lump in your throat that developed at the thought of so many bacteria. If we ever do find extraterrestrial life, chances are they will be microbes. (This is why one of the tasks of the Mars rovers is to search for signs of an environment capable of supporting microbial life.) At more than 3.5 billion years old, single-celled microbes are the oldest form of life on Earth. By comparison, humans emerged just 200,000 years ago. If you set the history of earth to a twenty-four-hour day, with the planet's creation occurring at midnight, microbes would have appeared a little after 4 a.m., while humans would have appeared only a few seconds before the end of the day. Without microbes, humans would not exist, but if we all disappeared, few of them would notice.

Despite their seemingly primitive forms, present-day microbes are the product of billions of years of evolution. These microbes are therefore just as evolved as we are—in fact, considering the many more generations microbes have gone through (they reproduce on the time scale of minutes to hours), you could argue that they are better adapted to the current environment than humans. For example, within just a few decades fungi able to harvest energy from radiation have become prevalent near the site of the Chernobyl disaster. Should widespread devastation strike the planet, certain microbes would likely be able to quickly adapt to the new environment and proliferate. Our human bodies, on the other hand, cannot adjust as readily.

Every newborn child represents a habitat of fresh real estate for microbes. Since microbes are so plentiful and have an amazing ability to rapidly acclimate to new environments, they immediately take up residence on every body on the planet, human or otherwise. They find homes on our skin, in our ears and mouths, and in every other orifice on our body, including the entire digestive system, where most of them live. Although the microbes that inhabit us were, in the beginning, just looking for food and shelter, over the course of our coevolution they have become a fundamental part of our biology.

THE BACTERIA-FILLED TUBE (AKA THE HUMAN)

The human body essentially is a highly elaborate tube that starts with the mouth and ends at the anus. The digestive tract, or gut, is the inside of the tube. As Mary Roach noted in her highly entertaining book *Gulp: Adventures on the Alimentary Canal*, in this way we are not that different from the earthworm. Food goes in one end of the tube, gets digested as it passes through the tube, and is then excreted as waste at the other end. Before you get depressed about how "unsophisticated" our digestive system is, remember that the two-opening tube was a major advance over earlier one-opening tubes. The hydra, a microscopic animal that lives in ponds, has only a mouth. That means that ingested food and excreted waste share the same opening. Now our "tube" doesn't seem so shabby, right?

Unlike the worm, our tube has an assortment of accoutrements that have evolved to nourish and protect it. To feed our tube, arms and hands serve to reach for and grab food. We have evolved legs and feet to help us move around and find more food. All of our senses and our highly complex brain can be thought of as "extras" to get more food for our tube, protect our tube from harm, and to procreate, thereby making more tubes. Additional tubes provide new habitats to be occupied with ever more bacteria.

Despite the tremendous impact our gut-residing microbes have on digestion, food travels most of the length of our digestive tract before encountering the bulk of these microbes. The food we ingest makes its way down the esophagus to the stomach, where it lands in a bath of acid and enzymes tasked with starting the process of digestion and nutrient extraction. After about three hours of mechanical churning in this harsh, acidic environment that is relatively devoid of microbes, the partially digested food is slowly emptied into the small intestine. This is where the digestive system truly begins to resemble a tube. This flexible passageway is approximately twenty-two to twenty-three feet long, an inch in diameter, and piled like a plate of spaghetti in the middle of our body. The interior of the small intestine is covered with finger-like projections called villi that absorb nutrients into our bloodstream.

The food traveling through the small intestine is soaked in enzymes secreted by the pancreas and liver to help digest the proteins, fats,

A scanning electron micrograph of villi within the mouse small intestine. © Justin Sonnenburg, Jaime Dant, Jeffrey Gordon

and carbohydrates we've consumed. Here in the small intestine the microbe count is relatively sparse, with *only* about 50 million bacteria per teaspoon of contents.

The last stop in this roughly fifty-hour journey is the large intestine, or colon, where food moves through at a snail's pace. The large intestine is not as long as the small intestine—less than five feet on average—but its name comes from its width, about four inches in diameter. A layer of slimy mucus coats the inside of the large intestine. It is here that the remainder of the food we've eaten first encounters the dense and voracious community of microbes referred to as the microbiota. (The large intestine contains about 10,000 times more bacteria per teaspoon than the small intestine.) Gut bacteria live, and in fact thrive, on leftovers, primarily the complex plant polysaccharides known as dietary fiber. Whatever the bacteria don't (or can't) consume, for example seeds or the outer skin of corn kernels, is excreted some 24 to 72 hours after the initial esophageal descent. Included in this waste are lots of bacteria, some dead and some still living, that get swept along with the current. Close to half of the mass of stool are bacteria, but they leave plenty of their brethren behind to ensure the tube remains densely populated. Depending on the existing sanitation standards, some surviving microbes may spread to a nearby water source, allowing them to find a new home in someone else's tube.

How did all these bacteria get into our digestive system in the first place? We often think of our insides as being, well, inside of us. The reality is that the inside of our tube is exposed to our external environment in the same way our skin is exposed on the outside of our body. That is the nature of a tube, after all. Through repeated exposure to the microbes that surround us—on our hands, in our food, and on our pets—our tube is constantly exposed to microbes. Some of them pass through us, but some of them stick around for years or even our entire lives.

Despite their prevalence in the colon, the life of a gut microbe is not easy. First they need to withstand the acid bath that is our stomach and then ultimately find shelter in the dark, damp cavern of the colon, which is inhabited by more than a thousand different species. While food periodically arrives in the cave, competition for resources within the gut is fierce and survival depends on snatching what you can before others get their microbial hands on it. In between meals, some microbes survive by dining on the layer of mucus that coats the intestine.

While life has always been a struggle for gut microbes, never has this been more the case than today, given what they are facing in the Western world.

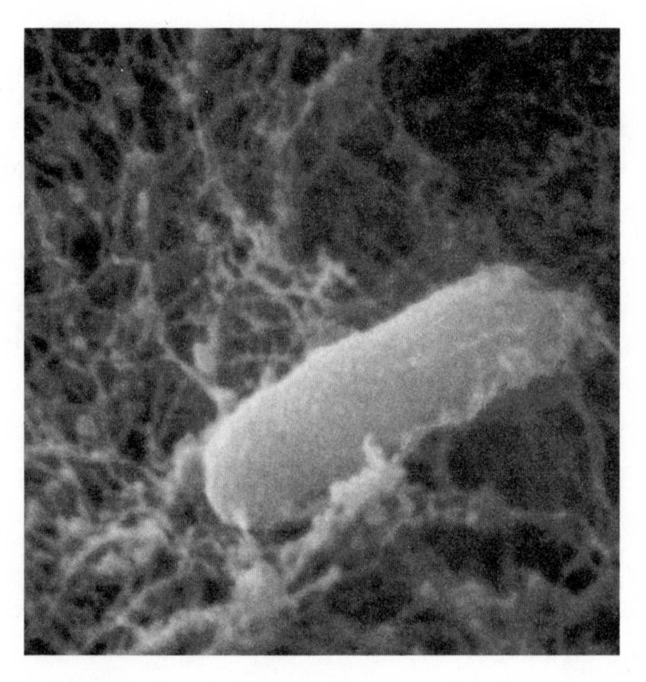

A scanning electron micrograph of a rod-shaped member of the human microbiota embedded within mucus. © Justin Sonnenburg, Jaime Dant, Jeffrey Gordon

THE WESTERN MICROBIOTA PLANE WRECKAGE

Imagine that your first image of an airplane was a picture of a debris field after a plane had crashed. Knowing nothing about aviation, you'd find it difficult to piece together what the airplane looked like before the crash. This analogy is akin to what researchers face when they try to understand how the human microbiota works. The vast majority of microbiota research has been performed on people from the United States or Europe—the same individuals who are predisposed to Western diseases. When scientists compare the microbiota of people with inflammatory bowel diseases (IBD) to those without it, they are cognizant that the "healthy" group, by living a Western lifestyle, may not provide an accurate definition of a healthy microbiota. One of the hazards of modern society is the increased risk for developing IBD. Although an individual may not have IBD yet, their microbiota could already be in an unhealthy state, tending toward illness in the relatively near future. It would be like comparing someone with a cold accompanied by a fever and a cough to someone who has a fever but hasn't yet developed the cough. In this scenario it could appear that having a fever is normal (even the "healthy" person has a fever) but that coughing is the problem. Because our definition of a healthy microbiota comes from studying Americans and Europeans, it's likely that our view of what is normal is highly distorted.

From the birth of humanity until about twelve thousand years ago (a time span of about two hundred thousand years), humans obtained their food exclusively through hunting and gathering. The ancient human diet consisted of sour, fibrous, wild plants and lean, gamey, wild meat, or fish. The birth of agriculture marked a dramatic change in the way people ate. Domesticated fruits and vegetables (selectively bred for increased sweetness and plumper, less fibrous flesh), grain-fed animals and animal products like dairy, and cultivated grains like rice and wheat became common fare for our species. Over the past four hundred years, the Industrial Revolution brought unprecedented and rapid change to our diet, which increasingly relied upon mass-produced food. Modern-day technology over the past fifty years has resulted in grocery stores filled with a seemingly endless supply of highly processed, overly sweetened, calorie-dense foods that have been stripped of dietary fiber and sanitized to prolong their shelf life. A diet filled with these new food products represents a huge deviation from what we have eaten over most of our evolutionary history. The gut microbiota has ridden this dietary roller coaster throughout human history, constantly adjusting to each shift in food technology and dietary patterns. Unfortunately, it now appears to be on a potentially disastrous trajectory.

One of the marvels of the gut microbiota is how rapidly it adjusts to dietary changes. The bacteria in the gut divide quickly, capable of doubling in number every thirty to forty minutes. Species that thrive on the types of food an individual regularly consumes can become very abundant relatively quickly. However, species that require food that are not a part of the person's normal diet can become marginalized, relegated to subsisting on intestinal mucus or, in the most extreme conditions, face extinction. In biology, this ability to change is known as plasticity, and the gut microbiota has it in spades. Microbiota plasticity ensured that when our ancestors' hunter-gatherer diets changed with the seasons, their microbiota could easily adjust to extract the maximal nutritional benefit. However, this plasticity also means that once-abundant species, well suited to a foraging diet, have now disappeared in the face of our modern diets. Conversely, microbes that thrive in today's burger-and-fries environment are becoming a larger proportion of the microbiota. This Western microbiota is the one most of us now harbor in our gut, even those of us who consider ourselves healthy; and unfortunately, the picture probably looks more like a crashed airplane than a fully functional one.

To get a sense of what a fully functional microbiota might look like, we can look to the last remaining full-time hunter-gatherers in Africa, the Hadza. They live in the cradle of human evolution, the Great Rift Valley of Tanzania, home to some of the most ancient remains of our human ancestors dating back millions of years. Their diet and microbiota provides the closest modern-day approximation to that of our ancestors who lived before the advent of agriculture.

The Hadza consume meat from hunted animals, berries, the fruit and seeds of the baobab tree, honey, and tubers—the underground storage organs of plants. The tubers they eat are so fibrous that after a period of chewing diners spit out a cud of the toughest fibers.

Those who have studied the Hadza estimate they consume between 100 to 150 grams of fiber per day. To put these numbers into context, Americans typically eat only 10 to 15 grams of fiber per day.

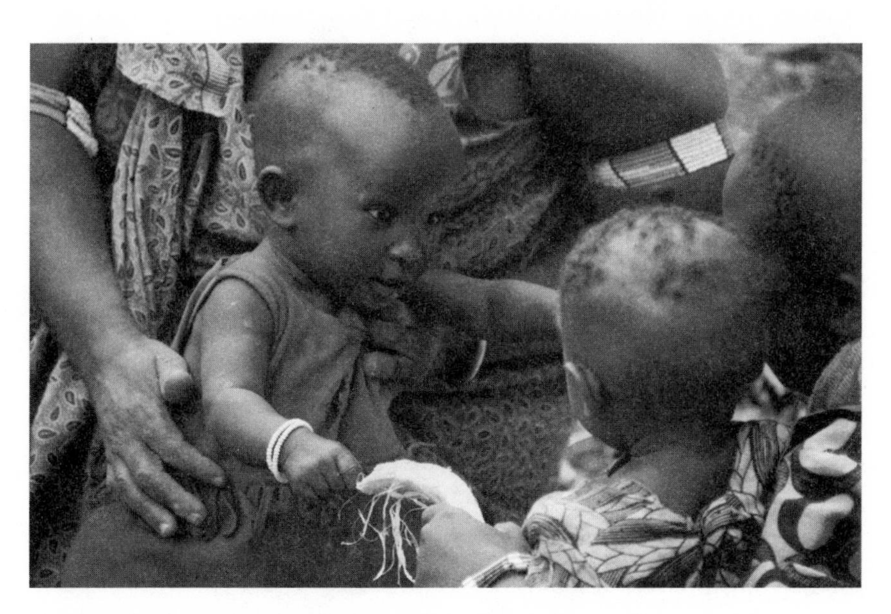

A young Hadza girl with a freshly cooked and peeled piece of Vigna frutescens *tuber.* © Pascal Gagneux

The Hadza microbiota houses a much greater diversity of microbes than a Western one does. If you think of the microbiota as a jar of jelly beans with the different flavors representing different species of bacteria, the "hunter-gatherer" microbiota is like a jar filled with a complex mixture of many different colors and flavors, some of which are very unusual. The jar representing the Western microbiota has far fewer flavors in a more homogenous or simple mix.

The microbiota from individuals living a traditional agrarian lifestyle similar to how humans lived ten thousand years ago also contains a more diverse collection of microbes than Westerners typically house. These Western versus traditional differences are not just confined to the microbiota of adults. Children living in a rural village in Burkina Faso and in the slums of Bangladesh also have a microbiota that looks different from that of their European and American counterparts. Similarly to what has been observed in adults, Western children have a less diverse collection of microbes in their gut compared with children living a less modern lifestyle. Evidence is thus mounting that the Western microbiota contains a less diverse collection of microbes compared with the microbiota of people who don't consume much, if any, processed foods, aren't prescribed multiple rounds of antibiotics annually, and don't carry hand sanitizer in their purses and backpacks.

Diversity matters. In an ecosystem like that of the gut, diversity can be a buffer against system collapse. Imagine an ecosystem that contains a large variety of insects and birds. If one species of insect disappears, the birds still have a selection (albeit smaller) of insects to feed on. If, however, more and more species of insects disappear, eventually the birds will starve, compounding the depletion of species within the ecosystem. As diversity is lost in the Western microbiota, this ecosystem is at greater risk of collapse—a collapse that could affect the health of the human hosting the flailing ecosystem.

THE FORCED PARTNERSHIP

Humans are the evolutionary product of a lineage of organisms that continually figured out how to play nice with their gut microbes. Because the colonization of our gut by microbes was inevitable, our body had to learn how to interact with them in a positive way. The harsh reality of natural selection is that humans and bacteria are locked into a relationship by force. We have no choice but to coexist with them, so by making this partnership positive, both humans and bacteria can benefit.

Although some species, such as *Salmonella*, *Vibrio cholera*, and *Clostridium difficile*, commonly referred to as pathogens, have taken the route of antagonistic interaction, these are the exceptions to the masses of friendly microbes that we harbor. Unfortunately, pathogens have driven the overuse of antibiotics to the detriment of the rest of the well-behaving members of the microbiota. If we cast our gut resident bacteria as invaders—or even as unimportant, as evidenced by our casual approach to antibiotic consumption—we risk harming this community and in the end harming ourselves.

Each species of microbe within your microbiota has its own genetic code, or genome. The collection of genes encoded within all microbes is called your microbiome, a second genome. Just as your human genome is uniquely yours (with the exception of identical siblings) no two gut microbiotas are identical. Therefore your microbiome is a major contributor to your individuality (especially if you have an identical twin). You can think of your microbiome as an internal fingerprint of sorts. Your microbiome may encode the ability to degrade a certain type of carbohydrate that someone else's microbiota can't. For example, some Japanese host a seaweed-consuming gut bacterium that is typically absent in the microbiota of Westerners. Because seaweed is such a large part of the Japanese diet, their microbiota has evolved a way to utilize this ubiquitous food source.

Hopefully the hallmark of the Western microbiota is not the ability to consume hot dogs!

We need our gut microbiota. Because humans had no choice but to house this dense collection of bacteria, we did what all evolutionarily successful organisms do: we entered into a mutually beneficial symbiotic alliance. In other words, we made them work for their keep. Symbiosis is defined as a close and extended relationship between two or more organisms. Some symbiotic relationships are parasitic, meaning one organism benefits at the expense of the other, like the unwanted houseguest who eats all your food, leaves a mess, and doesn't get the hint that it's time to go. On a microscopic level, hookworms are a great example of an unwanted houseguest. Commensalism is a second type of symbiotic relationship, one that benefits one participant but has little or no effect on the other—imagine a dog that raids your trash for food. In mutualism, a third kind of symbiosis, both parties benefit. Now imagine that the dog raiding your trash is also keeping disease-spreading rats away. This arrangement is analogous to the relationship we are in with our gut microbiota.

The most obvious way that we benefit from the microbiota is from the chemical products they release (and we absorb) during the fermentation reactions they carry out in the gut. These chemical reactions allow us to salvage calories from food that would otherwise be wasted, something that would have been critical to our ancient ancestors in their calorie-sparse environment. While extracting extra calories is less important in the modern world, these reaction products still perform important biological tasks for us: tuning our immune system, helping us fend off disease-causing bacteria, and regulating our metabolism.

Gut microbes receive a steady supply of food, provided by us, without having to expend much effort other than to wait for its appearance. So instead of "you scratch my back and I'll scratch yours," it's more like "you eat food for me and I'll help you digest it into molecules that you need." But why doesn't the human genome just encode the ability to completely digest our food so that we don't have to deal with these freeloading microbes? One reason our digestive tract is not free of microbes is because the task of microbe elimination would be nearly impossible to accomplish. Trying to maintain a microbe-free existence in our microbial world would be a herculean effort requiring our immune system to work around the clock to evict the plethora of microbes we are constantly encountering.

Another reason we don't eradicate all our microbes is because their genes function as an extension of our own genome. Each gene within the human genome provides a benefit, but also comes at an energetic cost. Every time a human cell divides, the genetic material from the entire human genome contained within that cell (roughly twenty-five thousand genes) needs to be replicated. We profit from microbial genes that perform a variety of functions that our genome cannot. For example, the microbial genomes provide the ability to convert otherwise indigestible food into key molecules that regulate many aspects of our biology, from the amount of inflammation in the gut to how efficiently we store extra calories. This coevolved division of labor is so successful that it has been used by organisms for eons.

Tremblaya princeps is a bacterium that lives inside of a garden pest known as the mealybug. This microbe is special because it has one of the smallest genomes of any bacteria presently known and represents a minimal number of genes required for life. Scientists are interested in small genomes because they provide a good starting point for engineering microbes from scratch to perform helpful tasks, such as cleaning up oil spills in the ocean or converting cornstalks into fuel. After the genome of *Tremblaya princeps* was sequenced, it became obvious that this bacterium was missing key genes required for even the most basic cell functions. Nested inside *Tremblaya princeps* is another bacterium called *Moranella endobia*, which contains the missing genes needed by *Tremblaya princeps*. *Tremblaya princeps* has utilized an extremely clever strategy: instead of maintaining all the genes it needs for life, it has co-opted the genes of another bacterium, *Moranella endobia*, allowing both bacteria to survive in a mutually beneficial way.

Nature long ago figured out what so many of us still struggle with today: the key to success in a competitive environment is to delegate and work together!

The relationship between *Tremblaya princeps* and *Moranella endobia* is highly analogous to our alliance with the gut microbiota: house bacteria, assign necessary functions to them, and maintain a more streamlined genome. The catch: you have to take care of the important bacteria that are performing vital functions.

We rely on the genes within our microbiome to make up for deficiencies we have in our own genome. Degrading the variety of dietary fibers found in the plant material we consume requires the vast collection of genes our gut microbes provide. Our symbiotic relationship with these bacteria throughout human history has made us reliant on the chemical cues they provide to the various systems within our body from birth to the end of our lives. These cues ensure that our intestinal tract develops properly after birth, that our immune system aggressively fights disease (but not overzealously), and that our metabolism maintains homeostasis. The human genome benefits from the 3 to 5 million genes the microbiota provides, without being taxed with maintaining them.

BACTERIA'S BAD REPUTATION

If the microbiota is so important to human health, why are we just hearing about it now? Until very recently, the field of medical microbiology focused on "bad" bacteria, also known as pathogens. These pathogenic bacteria are the cause of human diseases such as cholera, tuberculosis, and bacterial meningitis and have been responsible for the suffering and death of countless people throughout history. It is not hard to see why medical research has been so focused on understanding and fighting these bacteria. In the mid-1800s Louis Pasteur, the famed microbiologist, began experimenting with the idea that microorganisms were the cause of food spoilage and fermentation—the process that turns milk into yogurt or grape juice into wine. Before Pasteur, scientific thinking maintained that an entity of some kind spontaneously generated within milk to cause it to spoil. Through his experiments, Pasteur proved that spoilage and fermentation was not a result of a ghost-like apparition but must be the result of something within the environment. That something was microbes.

Pasteur's work helped to fuel the germ theory of disease. Pasteur believed that, just as microbes could spoil milk, human diseases could also be the product of invasions by microscopic organisms. The germ theory of disease was a highly novel concept at the time because the prevailing scientific thought was that human diseases were caused by miasmas—foul-smelling poisonous air from rotting organic matter. This belief in disease-causing miasmas drove decisions about sanitation.

The mid-1800s was a time of improved personal sanitation in London. The revolutionary invention of the flush toilet was embraced by households eager to get rid of their less than sanitary chamber pots. But new problems arose with the flushed waste. London did not have much of a sewer system at that point and chamber pots were mostly dumped into cesspits throughout the city. Flush toilets were designed to drain into these cesspits; however, the water that accompanied the flushing quickly resulted in overflows into the Thames River, a source of drinking water for many residents of London. By the mid-1800s deaths from cholera were rising—much like the sewage in the Thames.

The crisis came to a head in the unusually hot summer of 1858.

Between the high temperatures and the fermenting sewage in the Thames, London faced a huge smell problem referred to as the Great Stink. The odor was so overpowering that many people, in a kind of bacteria-induced house arrest, avoided leaving their homes. Because the increase in stench was accompanied by an increase in the incidence of cholera, it was easy to assume that the "miasma" stemming from the Thames was causing the cholera outbreaks. Cholera is a disease caused by the bacterium *Vibrio cholerae*, although this fact was unknown at the time. This bacterium is highly successful at spreading because the main symptom it causes, diarrhea, also disperses the microbe, especially in areas where sewage and drinking water mingle. In mid-nineteenth-century London, people obtaining their water downstream of the Thames were almost four times more likely to get cholera than people drinking farther upstream. The stench was just a sign of the lack of sanitation, not the cause of the cholera spreading, as people believed. But because sewage smell and ease of disease transmission are so linked, it was easy to believe that the noxious smell itself was the cause of disease.

Eventually, the Great Stink became so unbearable that people were forced to improve sanitation around the Thames. Cleaning up the water supply reduced the amount of *Vibrio cholerae*, thereby reducing the incidence of cholera.

The scene in London embodied centuries of civilization battling invisible foes, bacterial pathogens that caused pain, suffering, and death. But it wasn't until the 1880s that Robert Koch, a German scientist, demonstrated that bacteria were the causative agents for the diseases anthrax, cholera, and tuberculosis. His groundbreaking method, known as Koch's Postulates, still serves as the standard for establishing that a pathogen is the causative agent of a disease. Koch's findings earned him a Nobel Prize, as well as a position at the University of Berlin, where he served as director of the Institute of Hygiene—foreshadowing the coming century's obsession with cleanliness. His scientific discoveries finally put an end to the miasmatic theory and represented the birth of medical microbiology. For the ensuing 150 years, microbiologists focused on these disease-causing bacteria. Infectious disease represents the biggest killer of humans in our history. Antibiotics were developed as a direct result of our understanding that killing bacteria could stop many infectious illnesses and save lives. It is not hard to see how bacteria developed such a bad reputation and why societies moved toward increased sanitation.

It wasn't until the early 1900s that scientists began to appreciate the vast colony of bacteria living in the human gut. In one of the most understated discoveries of the twentieth century, Arthur Kendall remarked in the journal *Science* that "these experiments suggest that man has a bacterial population in his intestinal tract."

Despite knowing that bacteria were living inside of us, no one was quite sure what they were doing and if or how they were affecting our health. Pathogenic bacteria cause many acute human illnesses, while gut bacteria have a comparatively subtle, long-range impact on human health. As a result, research dollars were historically spent on the known "bad" bacteria. Only recently have scientists begun to grasp the enormity of the impact gut bacteria have on every aspect of our biology. To modify Arthur Kendall's observation, man doesn't just have a bacterial population in his intestinal tract, man is a product of this bacterial population.

THE COMING OF AGE OF THE MICROBIOTA

In the 1960s and '70s a group of visionary microbiologists, including Abigail Salyers, began studying bacteria from the human gut. One can only guess why these scientists chose to focus on these seemingly innocuous gut residents instead of the more flashy, disease-causing bacteria, but thankfully for us, gut bacteria piqued their interest. Salyers focused on a particular type of gut bacteria called the *Bacteroides* long before studies revealed that these bacteria are connected to human health in multiple respects. We visited her in her lab at the University of Illinois at Urbana-Champagne in 2005.

Abigail Salyers was a combination of fearless microbiota pioneer and pragmatic experimentalist. When we arrived she guided us through her lab and adjacent hallways that were packed with artifacts of her early microbiota experiments. Finally we settled in a meeting room. While seated around a table, we asked her why she chose the *Bacteroides* to study, hoping she'd offer up some special insight into her prescient choice. She replied that they were the easiest ones to work with because they could survive oxygen exposure. (Many other prominent gut microbes wither when outside of the oxygen-free environment of our gut.) One of her key discoveries revealed that this major group of gut bacteria was particularly adept at consuming dietary fiber. Salyers and her contemporaries laid the foundation for our understanding of how many bacterial species subsist in the gut—by eating the portions of plants that humans cannot digest on their own—but microbiota research at that time was limited by the tools available to researchers and by how difficult these bacteria are to work with in the lab. The field was waiting for new technology that could propel it forward.

The springboard for that leap was built in the late 1980s at the onset of the Human Genome Project. This international effort to sequence all the genes of the human genome was a massive undertaking. It took approximately thirteen years to finish sequencing the human genome and the total bill was about $1 billion. In the end we were left with about a terabyte worth of sequencing data that scientists are still working to unravel. The importance of this genomic leap to spur scientific discovery is irrefutable, but many feel that the completion of the human genome has not produced as many tangible benefits to human health as quickly as many had hoped, especially considering the price tag. Most will agree that the human genome is contributing to the development of important therapies and to our understanding of human disease, but the great promise of "personalized medicine"—treatments tailored to an individual's genome—has been slower to materialize than the pre-genome hype had pledged.

As benefits from the human genome slowly accrue, one unexpected and enormous outcome of the project was the incredible innovation in DNA sequencing technologies. Using present-day sequencing techniques—largely developed as a result of the Human Genome Project—the project could be finished in about a week and cost less than $5,000. In the near future, each of us will be able to have our complete personal genome sequenced in one day for around $1,000, largely due to the amazing technological innovation the project spurred.

While clearly a huge scientific and medical milestone, sequencing the human genome was followed by the growing realization that humans were much more than the product of their human genes. To obtain a full understanding of the genetic material we carry around, we also needed to sequence the genomes of our bacterial inhabitants, including those from the gut, but also from our skin, nasal passages, oral cavities, and urogenital tract. After the Human Genome Project ended, in 2008 the National Institutes of Health launched the Human Microbiome Project. The goal of this project was to characterize the bacterial life associated with the human body using the technology created for the Human Genome Project. Based on current estimations of the amount of bacterial genetic material humans are host to, we are only about one hundredth of the way to having all our associated genes sequenced. Now, about seven years after the onset of this program, scientists are starting to get a better understanding of the microbes that call our body home and it's paving the way for a new, more complete version of personalized medicine.

As technology progresses, the personal microbiome, although

more than a hundred times larger than our personal human genome, will provide each of us with a mind-boggling amount of information about the microbial communities we harbor. Now we are able to ask questions like: How does the microbiota change in individuals who have a specific disease? How do factors ranging from owning a dog to eating seaweed affect the microbiota? How fast can my microbiota shift after I've changed my diet?

Presently, microbiota research is going through a phase that is analogous to the Human Genome Project, with labs across the world using modern sequencing technology to obtain a census of our gut inhabitants. To complement the sequencing efforts, many laboratories are using additional cutting-edge technologies to move our understanding beyond the DNA sequence to examine many facets of microbiota biology, such as the identity of the sea of chemical compounds these microbes produce inside our bodies. In the next decade we will achieve a level of understanding about our relationship with the microbiota that is likely to impact how we can prevent and treat numerous diseases.

While scientific pioneers are charting this new frontier, they are cognizant of the mistakes made in overly optimistic forecasts of how the Human Genome Project would rapidly transform medicine. It is not difficult to envision a backlash if expectations become overinflated, and with a biological system as complex as the microbiota, translating discoveries to medical practice will take time. But exercising restraint in enthusiasm for the microbiota's health implications—a topic that captivates scientists and nonscientists alike—is a bit like parking a new Ferrari in your driveway for your child's sixteenth birthday and asking the dealer to send the keys along in a few years. Imagine a researcher telling a family with an autistic child, "Yes, we have found a link between your child's affliction and the gut microbiota, but we are pursuing this lead very cautiously and will get back to you in about a decade."

THE FORGOTTEN ORGAN

A little over ten years ago, when we first started studying the gut microbiota, it felt like we were investigating a new organ in the human body. In fact, the microbiota is often referred to as the forgotten organ. There was so little known about how the microbiota worked, what types of bacteria were present, and how this organ contributed to our health. And yet, simultaneously, there was so much potential for how the microbiota could ameliorate diseases.

As in all new fields of scientific research, there is a period of "stamp collecting" that needs to occur. Scientists have spent the past several years cataloging which types of bacteria live in our gut. The Human Microbiome Project, among other efforts around the world, played an important role in this descriptive phase and has laid a strong foundation for scientists to pursue a deeper understanding.

The easiest way to obtain a gut microbiota census is to collect a stool sample from someone. Since human feces are by dry weight 60 percent bacteria, a simple DNA extraction from less than a teaspoon of stool, followed by next-generation DNA sequencing, can tell us which types of bacteria your gut houses. When fecal bacteria are compared to bacteria coming straight from the colon (as collected during a colonoscopy) the two samples are very similar. But asking someone for a poop sample can be a little uncomfortable, especially since scientists are generally not known to be the most outgoing group around. Thus, many of the first microbiota census experiments saw scientists relying on themselves. That meant that every once in a while we would go home with a plastic Tupperware container and come back to the lab the next day with our "contribution" to the experiments. Either the stigma is gone or scientists have become bolder, or perhaps the burning curiosity of knowing what's living in your gut trumps the cultural taboo, but today, for just $99 (plus a small stool sample), you can get a list of the bacteria in your gut microbiota through the American Gut Project. The thousands that have participated so far illustrate that we've clearly come a long way.

In addition to next-generation sequencing, one of the great tools available to study the microbiota is the gnotobiotic mouse. Gnotobiotic means "known life." These mice have a gut microbiota that is completely defined and under the control of scientists; hence the life within them is known. Gnotobiotic mice can be colonized with a human microbiota, creating so-called humanized mice. Specific human donors can be chosen who suffer from diseases such as Crohn's disease, diabetes, IBD, or obesity. Some gnotobiotic mice are maintained germ free, meaning their gut has no bacteria at all. Through studying these germ-free mice, scientists have a greater appreciation for all the functions the microbiota performs. Some functions, such as helping to extract calories and balancing the immune system, were not too surprising. Others, such as their ability to affect moods and behaviors, were unexpected.

Germ-free mice are the product of two germ-free mouse parents, but at some point someone had to make the first germ-free mouse. To do this, a C-section was performed on a pregnant mouse and the uterus containing the pups was dipped into a light bleach solution to kill off any bacteria that might be tagging along. These pups could not be exposed to their mother since that would risk transferring bacteria, so each pup had to be hand-suckled by the scientist using a sterile container filled with sterile milk, giving new meaning to the term "surrogate parenting."

These sterile mice are only given food that has been heat and pressure sterilized, are provided with sterilized drinking water, sleep on sterilized bedding, and live inside a sterile plastic bubble that does not allow a single bacterium to enter. These sterile bubbles or isolators, as they are called, form a completely contained environment. Even the air that enters the isolators is filtered to minimize contamination risk. Unlike people suffering from severe combined immunodeficiency or "bubble boy disease," the gnotobiotic mice don't have a compromised immune system, although the lack of a microbiota does impact their immune system and it is not considered "normal" (more on that in Chapter 3). Periodically the sterility of the germ-free mice is confirmed, usually by verifying that a fecal sample does not contain any unintended bacteria. As you can imagine, maintaining mice under these conditions is a huge effort and expense. The slightest misstep, like mistakenly providing unsterilized water or allowing a breach in the air filter, can compromise an entire colony of mice, months of research, and thousands of research dollars. However, this extra effort has allowed scientists to ask questions about the microbiota that would be impossible to explore otherwise.

THE MICROBIOTA TAKES CENTER STAGE

Our mentor, Dr. Jeffrey Gordon, is a gastroenterologist by training but a scientist at heart and a microbiota visionary. Jeff's lab is filled with row upon row of plastic isolators set on top of steel carts. Each one houses a collection of gnotobiotic mice. Some mice have no microbiota at all (germ free), some have a normal mouse microbiota (conventional), and some have a human microbiota (humanized). Through caring for these mice, scientists soon realized that mice with no microbiota ate more food than their conventional (normal) microbiota counterparts, but actually weighed less. They also saw that obese mice housed a different collection of bacteria in their gut than the lean mice did. These observations provided the first clues that the bacteria in the gut and weight gain were linked—but how? Was obesity causing the microbiota to change or was the microbiota itself responsible for causing obesity?

This chicken-and-egg problem is common in scientific research and is often difficult to address. Often we can say with certainty only that two factors (the microbiota and obesity, in this case) are correlated or coincident, but not necessarily causally related. However, here is where the power of the gnotobotic mouse can really be seen. Jeff's team transplanted the microbiota from the obese mice into lean mice with no previous microbiota. Suddenly the lean mice with the obese microbiota began to gain weight, even though there had been no change in their diet or exercise habits! What these scientists had shown, to the surprise of many, was that the gut microbiota is enough to cause weight gain in an otherwise lean, healthy mouse.

These findings forced the scientific community to reframe our view of the gut microbes. Clearly the microbiota is not just a collection of innocuous bacteria loitering within our gut. These bacteria are capable of profoundly changing the biology of their host and may be a major contributor to one of the most alarming health issues in the Western world.

More recent research is showing that the microbiota-obesity link is just the tip of the iceberg. Dysbiosis, or microbial imbalance, is observed in people with a variety of health problems such as Crohn's disease, metabolic syndrome, colon cancer, and even autism. In fact, it is getting more and more difficult to find a health condition that has *not* been linked to aberrations within the microbiota. While in many of these cases we still don't know the extent to which the microbiota causes these illnesses, it is clear that we need to embrace a new way of thinking about ourselves. The collection of bacteria living within our gut is intimately linked to our health in ways we are just starting to understand. As scientific study into our microbiota marches on, we predict that these microbial inhabitants will have their hand in every aspect of our biology, from our cardiovascular health to our mental well-being.

HELPING THE MICROBIOTA BLOOM

While much research is still required to unravel all the ins and outs of how the microbiota functions, we feel there is enough solid scientific evidence to begin to make diet and lifestyle adjustments to optimize the health of the microbiota and thus our overall health. Within our own family, we have used our knowledge of the microbiota to make some big changes in how we live. Lessons from our lab and those of other microbiota scientists around the world have guided how we eat, what we pack in our kids' school lunches, how we clean our house, and how we spend our free time. There is already much information about how the state of our microbiota changes throughout the different phases of our lives, from birth to the golden years. By understanding how we first acquire our microbiota, how and what it eats, how it taps into our immune system as well as every other aspect of our biology, and what happens to it after a round of antibiotics, we can make informed choices that maximize the health and resiliency of this most important community of hitchhikers.

CHAPTER 2

Assembling Our Lifelong Community of Companions

Just prior to birth, humans are sterile, devoid of any microbial inhabitants—this is the only time in your life that you were a mere collection of human cells. But in the instant you leave the safety of your mother's womb, you begin your lifelong relationship with microorganisms. Upon entry into the birth canal you immediately start to assemble a more complex identity and your collection of human cells turns into the human-microbe superorganism you will be for the duration of your life. Just as a new island rising up out of the ocean presents a vacant landscape that is populated over time by flora and fauna, the newborn's body begs to be colonized by microorganisms—and there is a land rush for this vacant habitat.

OUR FIRST MICROBIOTA INHABITANTS

The human infant is born immature in many ways. Some people refer to the first three months of life as the fourth trimester, tightly swaddling their baby and playing as much white noise as possible to recreate the environment of the womb. Anyone who has spent time around a newborn can attest to the fact that these babies don't seem quite ready for life outside the mother. The infant digestive system is also in a state of incompleteness at birth. The slimy mucous layer that lines and protects the intestinal wall is thin and patchy, leaving the infant gut exposed to possible invasion by nefarious microbes. Mice with no microbiota have an extremely flimsy layer of mucus that quickly thickens up when they are exposed to gut bacteria. As with mice, when bacteria are introduced into a newborn, a complex symphony of human gene expression erupts. Among the resulting movements is the formation of a mucous layer that fully coats the intestinal lining and is optimized for viscosity and thickness to protect the integrity of the gut for the child's lifetime. Think of this layer as a gooey internal suit of armor, keeping bacteria at a safe distance from the infant's intestinal cells and minimizing the chance that rogue bacteria might try to infiltrate the intestinal wall, gain access to the bloodstream, and cause a systemic infection. Creating this barrier is a gigantic endeavor. Since our entire intestine can be up to thirty feet long, the surface area that needs to be coated in mucus could cover the entire floor of a two-thousand-square-foot house, although it would be a nightmare to vacuum. Encounters with early bacterial colonizers can set the stage for the effectiveness of this mucous layer and how the infant's immune system will respond not just to "good" bacteria but also to disease-causing, pathogenic bacteria, viruses, parasites, and even allergens. If the mucus barrier is formed improperly the result can be a compromised barrier that allows the invasion of bacteria or toxins.

Unlike the adult gut, the newborn's gut still contains oxygen from its life in the womb. A baby's first bacterial inhabitants are tasked with the job of consuming this remaining oxygen (while being able to tolerate the presence of oxygen themselves) in order to create an oxygen-free environment. These early bacterial colonizers till the land, so to speak, prepping the gut environment for a new crop of anaerobic bacteria that thrive in the absence of oxygen and become lifelong inhabitants. But which bacteria form the initial microbiota is determined by the way that child enters the world.

A child that passes through the birth canal is first introduced to bacteria from the mother's vaginal microbiota and from her anus. The mother's vagina often contains a large proportion of the bacteria *Lactobacillus*, an oxygen-tolerant group of microbes that are commonly found in the gut microbiota of vaginally delivered babies. After passing through the microbe-containing birth canal, the typical rear-facing presentation of human infants, combined with the compression of the mother's distal colon during delivery (think of a tube of toothpaste), exposes newborns to a faceful of mommy microbiota. While this may seem unhygienic, it is likely no evolutionary accident that our introduction into the microbial world would be accompanied by a healthy dose of mother-approved bacteria. She may not get to choose your friends or your spouse, but it turns out Mom has a huge say in your long-term bacterial companions. The fecal bacteria from the mother's gut clearly have a proven track record of shepherding a human to reproductive age, and it makes sense that these "pre-tested" gut microbes would be given the first chance at the land rush. Since the baby's microbiota looks more like its own mother's vaginal microbiota than that of another mother, it seems that, in addition to the mother providing half of her child's genes, she is also passing down her microbiota.

C-section babies have a very different first meeting with bacteria. Their first encounter with bacteria comes from skin—not exactly the way nature intended. Unlike in a vaginal delivery, which provides the newborn with microbes specific to his mother, children born by C-section don't receive skin microbes specific to the mother. Their very first bacterial community is not "inherited" the way the vaginal microbiota is in the child delivered naturally. Scientists don't yet understand why this is the case. It could be that exposure to other surfaces in the hospital or the skin microbiota from health-care workers also colonizes the infant, making the mother's contribution less obvious. The microbiota of C-section babies tends to contain more of a type of bacteria called *Proteobacteria* and less *Bifidobacteria* than that of vaginally delivered babies—a less than ideal collection, as you will learn later in the chapter. Because more than a third of deliveries in the United States are by C-section, it is more important than ever to understand how our first exposure to bacteria influences our microbiota and our health in the long term. There have been a slew of recent studies looking at C-section babies and their propensity for everything from obesity to allergy and asthma, celiac disease, and even cavities; and it is all bad news for those babies that miss out on the bacteria associated with a vaginal delivery. In many cases a C-section delivery is absolutely necessary in order to deliver a healthy baby and to protect the health of the mother. But now that we know the role that delivery method plays in microbiota assembly, we may need to consider practices that ensure that a baby's first bacterial encounter is the most beneficial.

Rob Knight is a professor at the University of California–San Diego and an expert at determining which bacteria live where. He is helping to lead the Earth Microbiome Project, an endeavor devoted to describing the microbial consortia across the globe, from the deepest oceans to the driest deserts. He is also making pretty good headway in identifying bacteria from all corners of the human body, and provides this service through the American Gut Project to anyone who wishes to know about their own microbiota. By obtaining a census of all human-associated bacteria, scientists can create a baseline identifying which microbes are commonly found where and use this baseline to determine whether certain microbes cause disease.

Knight and his wife recently had a child born by C-section. Being keenly aware of the differences that exist between the microbiota of C-section and vaginally delivered children, Rob and his wife took matters into their own hands. Using vaginal swabs from the mother, they inoculated their daughter at multiple body sites to ensure that she was exposed to the bacteria she would have encountered had she gone through the birth canal. While this approach may seem somewhat rogue, it is probably the next best thing to being born vaginally, as far as the microbiota is concerned. Although still far from the mainstream, it is not hard to imagine that this practice may accompany all C-section deliveries in the near future. However, before taking a similar action yourself, it is important to consult with a doctor who is knowledgeable about the specifics of your situation.

Both of our children were born by C-section at a time when we didn't know how the mode of delivery affects the early microbiota; if we had, we also would have considered the swab inoculation method. To add insult to injury, our first child was given antibiotics within hours of her delivery. Talk about a one-two punch to her developing microbiota. We mention this to illustrate that even when you have the best intentions and knowledge at hand there are certain circumstances that leave you with only bad choices for your microbiota. The one piece of information we did have at the time came from a study looking at supplementing probiotic bacteria for premature infants.

PREMATURE BIRTH: MICROBIOTA COLONIZATION INTERRUPTED

Infants born prematurely often have several medical issues to face. Depending on how early the birth occurred, they might have neurological problems, immature lungs, and be at increased risk for infection. Their gastrointestinal tract is also not fully prepared for life in the microbial world. Due to the immaturity of their gut, preemies are at increased risk of developing necrotizing enterocolitis, a devastating illness in which the infant's immune system mounts an overwhelming inflammatory response to the intestine, resulting in the death of portions of the infant's bowel tissue. Once the process of necrosis starts, it is often very difficult to save the life of the infant; 20 percent to 30 percent of neonates that suffer from necrotizing enterocolitis succumb to its effects. While it is unclear which event or events precipitate the onset of necrotizing enterocolitis, preterm infants that develop this disease have a gut microbiota that is distinct from infants that remain healthy. Premature infants already have a microbiota that is different from that of full-term infants, one that contains a less diverse collection of bacterial types. But premature infants that go on to develop necrotizing enterocolitis have fewer bacterial types compared to healthy preemies as well as more abundant bacteria that are often associated with a less healthy community. The "altered" microbiota of the necrotizing enterocolitis preemies was even evident up to three weeks before the onset of symptoms. Scientists began to wonder whether low microbiota diversity, along with the bloom of less than ideal species, was contributing to the development of necrotizing enterocolitis and, if so, could supplementing beneficial bacteria to these neonates protect them from developing this devastating complication of an early birth?

Premature infants given beneficial bacteria, such as those belonging to the *Lactobacillus* family, are much less likely to develop necrotizing enterocolitis than infants that receive no bacteria therapy. The exact reason why these bacteria are able to keep necrotizing enterocolitis at bay is still not completely worked out, but there are some clues about what role they may be playing. It seems that developmental completion of the gut and nascent education of the immune system require cues from bacteria. This is even necessary in full-term infants. But in preterm babies, because their gut and immune system are so much more underdeveloped, it may be that their gut has a more difficult time recruiting bacteria like the *Lactobacillus*. Bacteria like these provide signals needed by the gut and immune system to mature, fend off problematic bacteria, and keep inflammation under control. Also, beneficial bacteria can serve as placeholders within the gut, effectively preventing disease-causing bacteria from taking up residence. Having a good collection of "starter" bacteria early in life can have huge implications for health.

While our daughter was born at full term, knowledge of how important early gut colonizers can be planted a seed in our heads that providing probiotic bacteria might help to mitigate the less than ideal start to microbiota assembly that resulted from her C-section birth and subsequent antibiotic treatment. For the first two weeks after we brought her home, we sprinkled the contents of commercially available *Lactobacillus GG* capsules into her mouth. Obviously this was not a scientifically rigorous placebo-controlled study, so there is no way to know for sure what effect the *Lactobacillus* was having on her microbiota or health. However, anecdotally, she did not experience some of the acute issues that infants in her circumstance sometimes do, such as an oral yeast infection known as thrush.

Yeasts are not bacteria and thus are not directly targeted by antibiotics. But our bacterial inhabitants can help control the yeast population on and within our body just by taking up space, so to speak. One way to think about this is to imagine an Apple store the morning it releases the latest iPhone. Once the doors open, floods of people, many of whom have been waiting overnight, stream in, clamoring for the latest must-have tech item. At some point, physical space in the store becomes filled and no more people can enter. The space that microbes can inhabit on our bodies, like the space in an Apple store, is finite. Beneficial bacteria can serve to take up space and thereby limit the amount of yeast that can flourish. As in the study of preterm infants supplemented with beneficial bacteria, it is possible that filling space within the gut with benign bacteria excludes pathogenic bacteria and minimizes the risk of necrotizing enterocolitis. Giving the most beneficial bacteria first crack at the resources available within the gut could be an effective way to block microbes we do not want around. In the case of our daughter, her early exposure to antibiotics may have hindered the initial bacterial pioneers attempting to colonize her intestine. This situation could have provided a window of opportunity for a set of malevolent microbes to gain an advantage and populate her gut. By supplementing with beneficial microbes, we like to think that we tipped the balance in her favor as her early microbiota began to assemble.

PREGNANCY: A TIME OF MICROBIOTA CHANGE

If you have ever witnessed or experienced the nesting urge of an expectant mother, you are aware of the tremendous behavioral transformation that pregnancy can bring: painting and decorating to create the perfect nursery, neatly stacking newly laundered baby clothes, and spending endless hours meandering through stores, assembling an arsenal of baby swings, car seats, and bouncy chairs. An expectant mother's body is also undergoing preparations for her child's birth: loosening the pelvic joints to ease the baby's entrance into the world and starting the production of colostrum to ensure that a nutritious postbirth snack is immediately available. But there is another part of her body that is also preparing for the impending arrival—the microbiota.

We first met Ruth Ley when, like us, she was a postdoctoral researcher in Jeff Gordon's laboratory in St. Louis. Ruth isn't afraid to put on a pair of galoshes and jump into the stagnant water of a

Mexican swamp to figure out what makes a complex microbial ecosystem tick. Now an associate professor at Cornell University and focused on the microbiota, Ruth and her team of researchers decided to set their sights on understanding how the complex ecosystem within the gut responds to pregnancy, one of the largest physiological changes a woman can experience.

During pregnancy, a woman's body becomes an incubator, a gnotobiotic isolator of sorts, to nourish and protect its newly forming life. It made sense to Ruth that in the middle of all the changes that accompany pregnancy, the microbiota might be adapting as well. Ruth's team of scientists looked at the microbiota of ninety-one different pregnant women throughout their entire pregnancy. They gathered information about what these women were eating, whether they experienced gestational diabetes, and even tracked the microbiota of the babies for up to four years after birth. They found that, like so many other aspects of a woman's biology, the microbiota changed dramatically from the first trimester to the end of the pregnancy. By the end of the pregnancy there were fewer different types of bacteria than were found at the beginning. In other words, the composition of the microbiota was undergoing a simplification, becoming less diverse as the pregnancy progressed. In fact, the last-trimester microbiota resembled the microbiota of an obese individual.

To see what effect this "third trimester" microbiota had on its host, Ruth gave a group of normal nonpregnant mice a transplant of the so-called third trimester microbiota. Those mice gained more weight than mice that were harboring a "first trimester" microbiota even though both groups of mice were eating the same amount of food and were not pregnant. The collection of bacteria that made up the "third trimester" microbiota was able to extract more calories from the same amount of food and store those calories in the form of extra weight gain for the mice that housed these bacteria. From an evolutionary perspective, the ability to maximize calorie extraction would be highly beneficial to the mother and the developing infant. More calories from less food would reduce the burden on the mother to find more food at a time when her caloric needs, in order to nourish her growing child, are high.

But Ruth's research team also noticed that the so-called third trimester microbiota, in addition to causing more weight gain, had the hallmarks of one that could increase inflammation, a seemingly undesirable effect. Women in their third trimester carried more *Proteobacteria*, a type of bacteria that is enriched in people with gut inflammation and dysbiosis, and less *Faecalibacterium*, a type of bacteria that helps reduce inflammation. This inflammation-inducing microbiota shift at the end of pregnancy seems counterintuitive since these bacteria will be the child's first exposure to the microbial world. Why would Mom want her child's first set of bacterial companions to be types that are associated with inflammation?

To their surprise, when Ruth's team examined the microbiota of the newly born infants, they found that it more closely resembled the mother's "first trimester" microbiota than the inflammation-inducing "third trimester" microbiota. It's not entirely clear why this would be the case. Maybe the bacteria thriving in the third trimester gut aren't a good fit for the infant gut, so while the infant encounters them during birth, they don't persist. The bacteria from the first trimester, while lower in abundance, are still present at the end of the pregnancy. These microbes seem best able to flourish in the infant gut. It appears that the newborn's gut functions somewhat like a change sorter at birth: regardless of the assortment of bacteria going in, the infant gut selects which ones it will keep and which ones will pass through. Some of this sorting is surely a result of genetic factors the child is born with, but there is growing evidence that environmental factors are also critical in determining which types of bacteria bloom during infancy.

HUMAN MILK: THE INFANT MICROBIOTA GUIDE

The first several months of a baby's life are punctuated by shifts in microbiota membership. There are periods of microbial blooms during which certain species become very abundant and then, for unknown reasons, subsequently crash. In 2007 a study from Stanford University looked at the developing microbiota of fourteen infants from birth until their first birthday. Known as "succession" in ecological lingo, the order in which species arrive and flourish within an ecosystem is governed by a set of rules. These researchers were hoping to discover a set of guiding principles, a road map, to describe how an infant goes from no bacteria to a fully complete and complex microbiota. They found, however, that gut microbiota assembly appears to follow a chaotic process typified by tremendous instability and that it was unique for each of the fourteen babies. Only two children shared a somewhat similar microbiota profile over the course of this first year of life—the only twins in the study. Since fraternal twins have many genes in common, as well as a similar environment, it's difficult to say whether nature or nurture was the prevailing reason for the similarity of their microbiota.

The seemingly disorganized first year of microbiota assembly may reflect our poor understanding of the complex interactions that occur among different bacteria as they try to create a stable microbial ecosystem from scratch. With more years of infant microbiota research, universal themes may emerge that explain how bacteria form complex communities in this virgin environment. Clearly much remains to be learned about how humans acquire and maintain our internal microbial associates.

While there is apparent randomness in early microbiota acquisition, it is clear that nature has not left it all to chance. The first food consumed by most infants around the world is breast milk. Human milk has been engineered by the forces of evolution to provide the child with the best possible chance at survival. Mothers invest a tremendous amount of resources into the production of breast milk. Five hundred extra calories per day may be required to produce enough breast milk for one child. By comparison, pregnancy only requires an extra three hundred calories per day. The ingredient list for breast milk reads like a who's who of supernutrients. Rich in fats, protein, and carbohydrates and numerous other health-promoting compounds, human milk provides complete nutrition for the baby. It is replete with tailor-made antibodies and other immune system molecules that confer passive immunity to the infant while his immune system continues to develop. Human milk also contains a less-well-known special ingredient called human milk oligosaccharides, or HMOs.

HMOs are a collection of complex carbohydrates that are the third most abundant class of molecules in breast milk after fat and lactose. Their chemical structure is incredibly complex, so complex in fact that humans do not have the capacity to digest it. That's right, one of the main ingredients in breast milk is not digestible by the infant who's drinking it. Why would a mother put precious energy into making something her baby can't use? The answer is that HMOs aren't for feeding the baby, they're to provide sustenance for his microbiota. The microbiota, with its collection of 25 million genes, has the capacity to digest and extract energy from HMOs. The lactating mom is not just providing for her baby, she is also whipping up dinner for the 100 trillion bacterial guests her baby is hosting. And in the form of a diaper change, she has to clean up after them as well!

By no accident of nature, the bacteria that are the best nourished by HMOs, such as the *Bifidobacteria*, also appear to be those most likely to be found in the guts of healthy babies. But HMOs do more than just feed beneficial bacteria the baby needs at this stage; they also help to seed another type of beneficial bacteria, the *Bacteroides* that Abigail Salyers and others studied. *Bacteroides* have an amazing capacity to thrive off plant material. By giving the *Bacteroides* an early advantage, HMOs are also preparing the baby for life on solid food. HMOs serve as a conductor, orchestrating the major transition in microbiota development that occurs with the introduction of solid food. In all aspects of a child's life, a mother tries her best to guide her child to make the wisest choices to succeed in a world where so many events are out of her control. HMOs illustrate, on a molecular level, how a mother guides yet another process in her child's life, one that is impacted by external forces—microbiota assembly.

Mothers also provide living bacteria through breast milk, although it is not clear where these milk bacteria originate. Do they live inside the lactating mother's breast, where the milk is generated, comprising a separate milk microbiota? Or do they come from elsewhere in the mother's body, like the gut, passing to her breast and, ultimately, into the infant through the milk? What types of bacteria are being transmitted and what that means for the health of the child are questions that science presently does not have answers for. What is clear is that a mother's milk shepherds her child's microbiota assembly to ensure that the most beneficial community possible takes up residence.

These findings have made baby formula companies acutely aware of the huge deficiencies in their product when it comes to feeding the infant microbiota. With microbiota health in mind, some formula companies advertise new ingredients in "premium" formulas in an attempt to mimic the contents of human milk. One such added ingredient, GOS, or galacto-oligosaccharide, is a manufactured carbohydrate that falls far short of emulating HMOs both in chemical structure and in its effect on the microbiota. Some formulas have even added living probiotic bacteria. At present there is little data that these additives help to approximate the full effect that human milk has on the infant and his microbiota. And, as you might guess, these premium formulas also come at a premium cost. By definition

HMOs are human specific; no other animal makes a mix of carbohydrates exactly like them. HMOs, because of their chemical complexity, are prohibitively expensive and take an extended period of time to manufacture industrially. The types of probiotic bacteria chosen to add to formula are just a best guess, since which types are ideal for babies is still unknown. While attempts to optimize formula to better serve the microbiota may be an improvement over those previously used, they are the product of only fifty years of scientific research and nutritional engineering. By comparison, breast milk is a result of thousands of years of human evolution; and while humans are marvelous engineers, the good money lies with the forces of evolution that have created the best nutrition for a baby.

The American Academy of Pediatrics recommends exclusive breast-feeding for the first six months of life and continued feeding, in combination with solid food, for an additional six months. The World Health Organization recommends providing breast milk for infants for up to two years and beyond. But if that type of commitment is not realistic, then mothers are encouraged to provide any amount of breast milk possible. Even smaller quantities of HMOs and milk bacteria (not to mention the numerous other health-promoting goodies found in breast milk) are likely to help shepherd the microbiota through the turbulent first year. Since we know the infant microbiota is unstable before the child's first birthday, it is reasonable to assume that providing some breast milk during that first year would be beneficial. We chose to breast-feed both our children, and we know how difficult breast-feeding can be. Our second child was not a natural breast feeder and Erica struggled getting breast-feeding established with her. But knowing how important it was for her health, we were proactive, seeking help multiple times from a lactation specialist—something not covered by insurance, even though it probably saved them loads of money due to fewer doctors' visits!

It's clear that one of the mistakes our society has made is not doing enough to promote breast-feeding. Remember that any amount of breast milk you can provide your child can only help get his or her microbiota started out on the right trajectory.

THE COLICKY MICROBIOTA

Most new parents are overjoyed when their child is born. Particularly in anticipation of a first child, we often have a romantic view of what life will be like with the new baby. While cognizant of diaper changes by the dozens and late-night feedings, many parents envision leisurely stroller walks and frequent naps with plenty of cooing, giggles, and smiles. But for about a quarter of all infants, these happy moments may be few and far between, with much of their time occupied by incessant crying. These are the infants that have colic.

For a parent, having a colicky baby can seem like a frustrating and helpless situation. No amount of comforting seems to soothe these babies. There are countless books written about colic and a number of "treatments" ranging from homeopathic gripe water to a rectally inserted catheter designed to relieve gas called the Windi. These unorthodox remedies speak to the desperation that parents with a colicky baby feel.

There is mounting scientific evidence that the gut microbiota may play a role in the development and severity of infant colic. A group of scientists in the Netherlands led by Willem de Vos looked at the microbiota of twenty-four infants for their first hundred days of life; half the infants had colic, half did not. They found that the microbiota of the colicky infants was much less diverse than that of the noncolicky infants. Strikingly, the colicky babies had more of the *Proteobacteria* type of bacteria in their gut and less *Bifidobacteria* and *Lactobacillus* bacteria, reminiscent of the microbiota of C-section babies and babies that are formula fed. Our daughter, who was born by C-section and was given antibiotics for the first two days of her life, also suffered from colic. While we do not know what the composition of her microbiota looked like, we would guess, based on the circumstances, that she had a low-diversity collection of bacteria and probably too much *Proteobacteria* and not enough *Lactobacillus* or *Bifidobacteria*.

If we had known about the connection between colic and the microbiota at the time of our first child's birth, we would have continued giving her the *Lactobacillus* probiotic beyond the first two weeks of life for a couple more months to try to ease her discomfort. If the *Lactobacillus* probiotic hadn't helped her symptoms, we would have tried additional types of probiotic bacteria until we hit on one that provided relief. There are a variety of baby probiotics available today; if you find yourself with an infant experiencing colic, it may be worth discussing these options with your child's pediatrician. Providing HMOs through breast milk is another way to promote the growth of *Lactobacillus* and *Bifidobacteria* and potentially ease colic distress.

WEANING: AN OPPORTUNITY FOR LONG-TERM MICROBIOTA HEALTH

Infants start consuming solid food at around six months of age. Anyone who has changed a diaper during that transition period can attest to the fact that solid food elicits massive changes in the infant's digestive system. When an infant is weaned onto solid food, the microbiota membership undergoes a radical shift and begins to resemble an adult's microbiota. A case study following the journey of a single child's microbiota from birth until age two and a half perfectly illustrates the development of a microbiota from infancy to a more stable, almost adult-like state. Throughout the two-and-a-half-year study, more than sixty fecal samples were collected from an infant; detailed notes describing dietary changes and health events in the child's life were also kept. The most dramatic change to the microbiota occurred upon the first introduction of solid food, in this case peas. The child's first foray into consuming plant matter resulted in a burst of microbial diversity, with different types of bacteria suddenly sprouting within the baby's gut. On an intuitive level, this change in the microbiota made sense. New types of food provided novel energy sources for the gut bacteria and opened the door for new types of bacteria to thrive. But what was amazing was how fast these new bacteria appeared after the peas had been eaten—within a day! Performing what resembled a clairvoyant's trick, the microbiota was prepared for the appearance of peas long before anything green made its way down the hatch.

Samples taken before the introduction of solid food showed that the infant's microbiota had contained, albeit at low numbers, the bacteria best suited for solid food even while still on an exclusive breast milk diet. How is this possible? The microbiota has an informant, HMOs, the special microbiota food present in breast milk. HMOs provide enough sustenance to allow plant-degrading bacteria to subsist at low levels in the gut during the milk-only phase of an infant's diet. Then, when plant material makes its first appearance, the bacteria are already seeded and ready to blossom.

Weaning is one of the most dramatic rearrangement periods for the microbiota in a person's life. Because this malleability comes in response to dietary change, it makes sense to wean children onto food that will maximize the health of their microbiota. We started our first child on solid foods the way many parents do, first introducing vegetables like peas, carrots, broccoli (all pureed, of course), then moving on to fruits. This order of vegetables before fruits followed the reasoning that if babies have too much fruit first they won't like the less sweet vegetables. Along with veggies and fruits, we fed our daughter rice cereal, oatmeal, and other types of grain cereals, dairy products, and meat. As she got older we started giving her portions of our own food instead of buying commercially available baby food, and always avoided food listed on "kids" menus. Our thinking was that we wanted her to develop a taste for "real" food, not kids' food like mac and cheese and chicken nuggets. In many cultures around the world, a baby's first foods are often just mashed versions of what the adults eat. From rice and lentils with a mixture of spices in India to hummus in the Middle East and even seal blubber in the Arctic, this method ensures that babies develop a taste early on for the food they will be consuming as adults.

The plan to feed our daughter a variation of our own diet hit a snag when, at the age of three, she began having issues with constipation. The issue became so bad that most trips to the bathroom ended in tears over her painful bowel movements. Her problem forced us examine what we were feeding her and, by extension, what we were eating ourselves. We thoroughly cataloged the variety and types of food we regularly consumed—a diligence that was in large part motivated by the fact that the microbiota and gastrointestinal health is our field of expertise. We felt that we, of all people, should not have a child with gastrointestinal issues.

To our surprise, we found that our diet was fairly homogenous and somewhat low in dietary fiber. Detailed cataloging was critical since before this exercise we would have said that our diet was full of fruits, vegetables, and whole grains. But we had been on autopilot, picking a limited variety of familiar foods containing refined, white-flour products, cheese, and a minimum of vegetables. When we were in the throes of being new parents, spending most of our days working and our nights not sleeping, there were few precious quality hours with our child. Providing a dinner that made for happy interactions, smiles, and minimal resistance was a highly attractive option. There was something innately satisfying about watching our two-year-old gobble up her dinner with a smile, even if that dinner was white pasta smothered in cheese sauce or mozzarella cheese melted in a folded white-flour tortilla. It struck us that our food choices were often fairly passive—what was readily available and inexpensive—and often our decisions were trumped by innate urges of hunger and tiredness: what could we put on the table quickly that would please even a finicky toddler's taste?

We decided to perform a major overhaul of our diet. We diligently, almost obsessively, observed the amount and types of dietary fiber we were consuming. Bags of white rice, white flour, white pasta, and pretty much everything that came in colorful packaging—often a marker of low-quality food items—all were tossed from our pantry. Empty shelves were filled with ancient grains like quinoa and millet, wild rice, and a variety of legumes. Our vegetable consumption increased to the point where the crisper drawers in our fridge couldn't hold our weekly allotment. We did not completely eliminate meat, but we frequently made legumes such as beans and lentils our main source of protein. Within days of these diet changes, which included much more dietary fiber from plants, our daughter's constipation disappeared and has never returned. Our second daughter, born after our newly reformed diet, has been free of motility issues from the start. This experience taught us a valuable lesson: feeding your children what you eat only makes sense if you eat a healthy diet yourself. Eating for health and a robust microbiota needs to be a family endeavor.

We believed that our first daughter's C-section birth likely seeded the wrong community of microbes, largely derived from skin, and that the subsequent regimen of antibiotics further degraded this community. Coupled with our less than ideal diet, these factors may have contributed to her persistent constipation. Without our dietary intervention these issues could have compounded, resulting in more serious problems such as irritable bowel syndrome or inflammatory bowel disease as she got older. Because we were so concerned that an unhealthy microbiota would put her on track for a lifetime of health issues, we made adherence to this plant-based diet nonnegotiable for our family. Before our diet change, persuading a toddler to eat steamed vegetables, or fighting with her at the dinner table, seemed like an exhausting, miserable undertaking. But we rapidly saw that, despite the initial hard work, using mealtimes as an opportunity to inform our children about the benefits of healthy eating was worth the effort.

In fact, we approach the challenge of having our children eat healthy as a combination of education and indoctrination. They are rarely thrilled to see steamed broccoli on their plates, but over the meal we discuss the importance of becoming "big and strong," of staying healthy, and of avoiding illness. We discuss the microbes in their colon that keep them healthy and that, in turn, are "counting on them" to send some veggies down their way. Five years into this process, we occasionally cave in to their will, allowing the occasional treat and letting a particularly unpopular vegetable go uneaten, but rarely. It is hard work, but we have learned several tricks to help in the process. A healthy dessert, for example a small piece of dark chocolate, can provide powerful motivation to finish a serving of lentil soup. Like so many other things that call for the indoctrination of children, such as religious beliefs, cultural values, and societal norms, we practice a sort of brainwashing of our children, insisting that eating healthy food is the only option. In our culture, we gather with family on a Thursday at the end of each November to eat a feast, stand during the national anthem before baseball games, and when children lose a tooth, put it under their pillow for the tooth fairy. Just like these other accepted practices, eating food that supports the health of our microbiota is just something that our family does. Junk food is never an available option in our house. And because as parents we eat the same foods we offer our children, we are continually modeling good eating habits, walking the walk, so to speak.

Today our children are six and nine years old. When we ask them why they eat vegetables they reply, "because they taste good." We have completely instilled our healthy-food "religion" in them and now they don't even bat an eye at the kale salad we serve at dinner.

If you are concerned that your child is too picky and that incorporating more dietary plants will never work, think about this: there are children around the world who eat insects, animal entrails, and numerous other items that we Westerners would find disgusting. They do it because these foods are part of their culture and because other choices are limited. If we asked them how they can stomach that stuff, we wouldn't be surprised if they said "because it tastes good."

ATTACKING A DEVELOPING COMMUNITY

Treating your child with antibiotics seems like a rite of passage for parents of young children in the Western world. And while the number of prescriptions for antibiotics written for children is decreasing, many feel it is still unnecessarily high. Antibiotics kill bacteria. The gut microbiota, being largely composed of bacteria, suffers tremendous collateral damage with each round of antibiotics. This unintended killing of our resident microbes, a version of friendly fire, has disastrous consequences on short-term and long-term health.

The infant microbiota case study showed that the introduction of solid food was followed by a bloom in microbiota diversity. But the opposite trend in diversity was seen after antibiotics. After a first round of antibiotics, the child's gut microbiota diversity decreased. This observation was not unexpected. Most antibiotics are broad-spectrum, meaning they are designed to kill many types of bacteria. In addition to the illness-causing bacteria, antibiotics also attack the "good" bacteria that form our microbiota. When the child had to repeat a course of the same antibiotic for a second infection a few weeks later, the diversity of the gut microbiota didn't take as big a hit as it had previously; the bacteria in this child's microbiota had adapted to endure the second round of attack.

The study demonstrates the two important ways that antibiotics impact your microbiota. First, antibiotics cause a profound and immediate decimation of the gut community. Second, even though microbes typically recover after the treatment ends, the microbiota, as a community or ecosystem, will likely never be exactly the same. After just one course of antibiotics the gut microbiota adapts. In the case of this infant, it became more resistant to attack when that same antibiotic was used again. Whether this adaptation is short-lived or a permanent feature of that child's microbiota for the reminder of their life is unknown. But current evidence suggests that this recovery is not perfect. Because the microbiota is wired into aspects of how the immune system functions (more on that in the next chapter), changes to this community have the potential to cascade into larger problems. Antibiotic use in children is associated with an increased risk for a number of ailments such as asthma, eczema, and even obesity. How antibiotic use and its subsequent effects on the microbiota can lead to these illnesses is not yet clear, but it appears that disturbances to the microbiota can lead to problems that on the surface seem unrelated to the gut.

WEIGHING DOWN OUR MICROBIOTA

Farmers have known for decades that giving low doses of antibiotics to livestock such as cattle, sheep, chickens, and pigs can increase their weight by up to 15 percent. Because meat is sold by the pound, increasing weight can mean extra profit for these farmers. The earlier in life an animal is provided with antibiotics, the more weight gain is achieved. The American child is prescribed, on average, one course of antibiotics per year. This fact has led scientists to wonder whether it is possible that frequent early exposure to antibiotics in our children results in increased weight gain. Might we be fattening up our children, inadvertently using the same method that farmers use to fatten up their cattle, with each antibiotic prescription?

Laboratory mice given low doses of antibiotics early in life, like farm animals, increased their percentage of body fat. Along with this ability to gain weight more easily, these mice have a microbiota that resembles that of an obese human and is different from that of a lean human. The antibiotic-treated mice ate the same number of calories as the nontreated mice but were better able to extract and store those calories in the form of added weight. A calorie uptake system that is dependent on the composition of the microbiota would explain the weight gain seen in antibiotic-treated farm animals and may even explain why antibiotic use in children has mirrored rising obesity rates.

A comparison between more than eleven thousand British children who either had or had not taken antibiotics found striking differences in body weight. Children who received antibiotics very early in life (before six months of age) weighed, on average, more than children of the same age who did not take antibiotics. This difference in weight was still observed up to three years of age. Children older than six months old who were prescribed antibiotics were also more likely to outweigh untreated children, although the effect was not as extreme as it was with the children who received antibiotics earlier. Children receiving antibiotics between the ages of one and two years old were significantly heavier than age-matched controls a full five to six years after treatment. These studies illustrate that antibiotic use early in life can have a direct impact on the composition of the microbiota and, what's more disturbing, on long-lasting weight gain and adiposity years after the antibiotic onslaught.

MICROBIOTA LESSONS FROM THE BEGINNING OF LIFE

When thinking about the microbiota at the beginning of life, there are five lessons to keep in mind that will help this community have the best possible start. First, as stated earlier, the method of birth matters. A vaginal delivery exposes the infant to the set of bacteria that nature intended. However, if a vaginal birth is not possible, as was the case with our two children, you can discuss with your doctor the possibility of exposing your newborn to a vaginal swab from the mother. Second, providing probiotics can serve as a buffer against factors, such as a premature birth, that result in a less than ideal first microbiota community. There are a number of commercially available probiotics that are formulated with babies in mind. Probiotics might also be indicated if your baby is colicky or has recently been prescribed antibiotics. Before starting your child on a probiotic regimen, it is important to discuss with your child's pediatrician whether probiotics would be appropriate and determine which types of probiotics are the best for your situation. Unfortunately, due to the individual nature of each person's microbiota, trial and error is often required to find a strain of probiotic that works best. We will elaborate on how to accomplish this in upcoming chapters.

A third way to help guide an infant's microbiota is through breast-feeding. Regardless of the method of birth, breast-feeding offers a great opportunity to provide both "mother-approved" prebiotics and probiotics to your infant. In the case of our second child who was born by C-section but not put on antibiotics, we felt that providing breast milk would be sufficient to help mitigate the less than ideal start to her microbiota. If exclusive breast-feeding is not possible, you may want to talk to your child's pediatrician about using a formula that contains prebiotics and/or probiotics. But keep in mind that any amount of breast milk helps, even if it's just nursing your baby at night before bed. Your milk, and the HMOs it contains, will help nurture a healthy microbiota within your growing child.

There are periods in a child's life when antibiotics cannot be avoided. However, a fourth beginning-of-life microbiota lesson to remember is how profoundly antibiotics affect the microbiota. Our new understanding of how antibiotics impact the microbiota and in turn our children's health means that minimizing their impact on the gut is critical. Providing breast milk is a great way to minimize antibiotic collateral damage. Supplying microbiota food in the form of HMOs as well as bacteria found in breast milk can help reseed the gut after the antibiotic storm has passed. In the absence of breast milk, formula that contains prebiotics and probiotics may help the infant gut recover. For children who have been weaned onto solid food, consider discussing the use of probiotic supplements, yogurt, or other fermented foods with the doctor prescribing the antibiotics. Whether probiotics help to counter the long-term health consequences of antibiotic use is not currently known; however, probiotics can help protect infants from pathogen-induced diarrhea, a common side effect of antibiotic use.

Finally, and in many ways most important, weaning offers a great opportunity to instill lifelong healthy eating habits in children. Habits that maintain a healthy gut community can benefit a child's health for his or her entire life. Getting kids to eat right can be a war of attrition. Be resolute in providing healthy choices, even if your child complains or refuses to try them. It often takes several attempts before a child is willing to try something new and even longer until the child enjoys eating a new food. The key is to not give up and succumb to nonhealthy choices. One approach that has been successful in our family is to explain to our children that they are keepers of a life form that lives inside of them, the microbiota, which needs to be cared for. We explain that the bacteria in our gut get hungry and while some of the food we eat is for ourselves, we also need to provide food for our microbiota. When explained in these terms, our kids are much more willing to finish their plate of vegetables—they feel they are helping the "pet" that lives inside them. In the final chapter we will discuss in detail what types of food are best for the microbiota and are also appealing to kids. The nurturing of the microbiota begins at birth and the better you can start things off, the easier it will be for your child to support a healthy microbiota, and maintain good health, throughout his or her life.

CHAPTER 3

Setting the Dial on the Immune System

CLEANING OURSELVES SICK

The Western world has seen a sharp rise in allergies and autoimmune diseases in the past half century. Many of us living in industrialized societies have experienced or have had a relative experience one of the many diseases that fit into this category—seasonal allergies, eczema, dermatitis, Crohn's disease, ulcerative colitis, and multiple sclerosis, to name just a few.

Why are these immune-system-related diseases so common now? Theories abound. Some blame our increased exposure to toxic chemicals and pollution or the chronic stress and depression we endure compared to the periodic intense stress our ancestors faced. Undoubtedly each of the diseases mentioned is complex and many environmental factors contribute to their onset, even within a single individual. Increasingly, however, evidence points toward the interplay between the microbiota and the immune system as central to the development of these diseases.

THE GUT: MISSION CONTROL OF OUR IMMUNE SYSTEM

Compared to other microbial environments on and in the body, for example on our skin or in our mouth, the relationship between the microbes within the gut and the immune system seems to be special. Our gut microbes are in constant communication with the part of the immune system located in the intestine. These microbe–immune system "conversations" help our body discriminate between harmless foreign entities like food and harmful ones like *Salmonella*. Clearly your immune system needs to respond differently depending on whether you've consumed a peanut versus a piece of contaminated chicken, and the microbiota helps train the immune system to make the distinction. But the microbiota's effect on the immune system response is not confined to the gut. Our systemic immune system—the immune system that circulates throughout our entire body—is also being instructed through its communication with the microbiota.

Because the gut is exposed to the outside environment—we are just a tube, after all—it is highly vulnerable to attack by outside invaders. For many pathogens the gut can serve as an entryway into our bloodstream and from there to other organs. But the body uses the gut's vulnerability and continual exposure to the environment to its advantage.

The immune system is highly mobile. Immune cells living in the intestine and "conversing" with gut microbes can suddenly pick up and leave the gut, enter circulation, and locate to new sites throughout the body. A T cell, one of the major classes of immune cells found in the body, living in your intestine today may be in your lung or spinal fluid tomorrow. And that cell can remember its experiences with the microbes in the gut. While this mobility may seem strange, from the perspective of human survival, it actually makes sense. Say a particular T cell encounters an invading pathogen while spending time in the intestine. It can multiply into many cells and spread throughout the body to inform other tissue of the impending danger. If that pathogen crops up in the lungs, an educated T cell is ready and waiting to help fight that infection. The immune system cells residing in the gut serve as important sentinels, alerted as soon as a possible invader appears and then coordinating a response from the gut throughout the body to prepare for a possible widespread scuffle.

You can think of the gut microbiota as operating a dial that controls the sensitivity or responsiveness of the entire immune system. Microbes in the gut can dictate local, gut-confined immune responses, such as how long traveler's diarrhea lasts, but also impact how your child will respond to a particular vaccine or how bad your hay fever will be this year.

The central role of the gut in properly preparing the body's immune defenses also means that the gut microbiota can, in some cases, misdirect the immune system. When interactions between the immune system and gut microbes are less than optimal, health throughout the body is negatively affected. If messages at mission control are misinterpreted, the immune system may respond too fast and too vigorously. And if the immune system is directed by the gut to operate on a hair trigger, autoimmune responses can result in which T cells and other immune cells take action against harmless things in the body.

A vivid example of how gut microbes can influence the trajectory of autoimmune disease was provided in 2011 by a lab at the California Institute of Technology in Pasadena. Sarkis Mazmanian led a group that was interested in how microbes in the gut could influence multiple sclerosis, a central nervous system disease that appears to have no connection to the gut. Performing experiments in mice, Mazmanian and his team demonstrated that the severity of an autoimmune attack on one's own nervous system could be dictated by which types of bacteria inhabit the gut.

The Caltech study is one of many demonstrating that gut microbes control the dial for how our immune system responds to perceived threats all over our bodies. Immunologists, who traditionally didn't think of gut microbes as anything more than an element in food's transition to feces, are now taking note of the microbiota and realizing that it is impossible to study and understand the most basic facets of our body's immunological responses without taking into account how far our resident microbes have turned the immune system dial up or down.

GUT MICROBES: PUPPETEER OF IMMUNE RESPONSE

The immune system is often described using battlefield lingo. When an infectious germ invades our body an army of immune cells and other molecules are mobilized to fight back and prevail over the opposing microbe. If, for example, you eat an undercooked piece of chicken laden with *Salmonella*, those pathogenic bacteria travel through your digestive system, where they can penetrate the cells that line your intestine. The cells of this lining release a storm of molecules called cytokines, which form what amounts to a molecular SOS to your body's immune system. Immune cells rapidly reply to the cry for help, homing in on the site of the invasion to confront the enemy. Ultimately, B cells and T cells, the foot soldiers of the immune system, work with numerous other specialized infection-fighting cells to rid your body of the offending intruders.

Meanwhile, as the host of this internal skirmish, you feel the fever and achiness set in, and as is often the case with a *Salmonella* infection, the need to run for the toilet. Described in these terms, it's easy to get the mental image of a raging battle every time our body encounters a microbe. Attacking invaders is an irrefutably important part of our immune system's job. Over the past several decades it is the view of the immune system as a heavily armed military force that has largely guided scientific research in the field of immunology.

But recent breakthroughs in the understanding of the microbiota have upended this simplistic model. With the increasing recognition of the trillions of microbes that reside in and on our bodies, immunologists have begun to embrace the much more numerous—in fact continual—interactions that our immune system has with the microbiota. The immune system is not just an army poised to do battle at the first sign of assault; it has a very substantial State Department as well. If the response to infections is symbolized by a country's readiness for war, the interactions with symbiotic microbes can be thought of as a government's ongoing diplomatic efforts. Much as in global politics, these more peaceful efforts by the immune system are a daily effort, and the battles much less frequent during times of crisis.

Our immune system is engaged in constant negotiations with our resident microbes over their shared resource—us. The immune system wants to enforce a safe distance between our human cells and our associated microbes. The microbes want to ensure access to their habitat—our gut—and not be expelled. The continual push and pull of these interactions can change in intensity depending upon what food you've been eating, whether you've introduced food-borne microbes, and many other factors. If "pushy" bacteria come to dominate your gut for a period of time, your immune system may be put on heightened alert. Typically, tensions will ease and our human cells and our microbial cells will reach something similar to a détente. However, during the time that relations became strained, your immune system, either within your gut or elsewhere in your body, will react very differently to a true assault by an invader. Like a race car revving its engine, the immune system is poised to zip into action, mounting a rapid and vigorous response. In cases where détente is not reached, the immune system can percolate to a heightened state of readiness, making it more likely to overreact to threats that are only perceived and not real. The results of this overreaction can range from a minor allergy to a painfully ulcerated colon.

Because the gut connects to the rest of the body's immune system, our resident microbes shape our immune responses in a global sense. The resulting decisions by the immune system, in turn, shape how it responds to pathogen invasion within the gut or at other body sites, how autoimmune diseases develop and progress, and which types of microbes are selected for eradication or allowed to remain within our microbiota. Some have suggested that the immune system should be renamed to reflect its true role as what might be called a "microbe interaction system." Its job in protecting us from harmful microbes is certain, but more frequent is the dialogue it maintains with the microbes we encounter on a daily basis. What happens if these conversations become less numerous? As the prevalence of immune-mediated diseases in the modern world rises, a tantalizing idea may explain this immune system dysfunction: we're too clean.

EVOLUTION OF THE HYGIENE HYPOTHESIS

In 1989, David Strachan, now a professor of epidemiology at St. George's University of London, proposed the hygiene hypothesis, which posited that the rise of hay fever and atopy (skin allergy) in the industrial world was a result of the reduced exposure to infectious agents. He suggested that the human immune system evolved in an environment in which it was constantly fighting off the plethora of disease-causing microbes encountered in food, water, and the overall environment on a daily basis. Hundreds of years ago, or even today in less modernized (traditional) societies, the human immune system had a full-time job ridding the body of the never-ending onslaught of disease-causing microbes. But now, thanks to antibiotics, sanitized drinking water, and sterilized food, we encounter far fewer microbes, reducing our immune system's workload to part-time. Based on the original observation that allergies were less prevalent in children with many siblings, the hygiene hypothesis suggested that children of large families were exposed to more sickness within their household, and so their immune system was occupied fighting infections and didn't have "time" to overreact to pollen or gluten and cause problems.

The hygiene hypothesis has progressed to incorporate the finding that children raised on farms have fewer allergies than children living in very clean, affluent households. It's not just exposure to disease-causing microbes but *any* microbes, like those found on farm animals or in dirt, that can occupy the immune system in a productive way. Although there is still ongoing debate about the complex interplay of factors and mechanisms behind the hygiene hypothesis, it's clear that the prevalence of autoimmune diseases in a population tracks with how effectively that population reduces its exposure to microbes. Sanitizing our environment, and eradicating microbes with antibiotics, has been incredibly successful in reducing the incidence of infectious diseases in our society. Unfortunately, the untargeted attack on disease-causing microbes has inflicted much collateral damage to the beneficial microbes caught in the crossfire.

Does this mean we need to be sick more often to ensure our immune system doesn't overreact? The answer seems to be no. The rise of autoimmune diseases appears to be more tightly tied to our increase in cleanliness, not to decreased infection. The vast majority of microbes we encounter are not destined to cause disease, but they do tickle the immune system in different ways; it continually revs its engine, albeit in largely unnoticed ways, at microbes that land upon, transit through, or live within us. These mild mini immune responses depend upon regular interactions with microbes and are part of maintaining a healthy immune system.

As our environment and food become more sanitized, we lose many of the microbial exposures we need to occupy our immune system. Antibacterial soaps and alcohol-based sanitizers seem to be spreading faster than the germs they are designed to combat. Kids carry cartoon-character-clad versions that hang on their backpacks and lunch bags. Grocery stores mount them on posts outside their doors like anti-germ security guards. As if slathering sanitizers on the outside of everything isn't enough, antibacterial chemicals like triclosan impregnate many pieces of kitchen equipment and even shopping carts and toothbrushes. There is even an antibacterial ice cream scoop, which makes us wonder what our risk of infection is with a regular scoop! Triclosan exposure has recently been linked to allergic responses and its prevalence is a marker of how obsessed our society has become with sterilizing all aspects of our lives.

The Western lifestyle is also further separating us from the soil microbes we once encountered in nature while tending crops or foraging for food. To make matters worse, the prevalence of antibiotics and antibacterial chemicals is not only limiting our exposure to harmless bacteria, it is increasing the prevalence of microbes resistant to these chemicals. Increased exposure to dangerous superbugs, such as those found in a hospital or in factory-prepared ground beef, sets off a spiral that only compounds the problem. Every story that hits the news about illness due to the contamination of cantaloupes, salad mixes, or hamburgers perpetuates our drive for microbial eradication, which may be fueling the rise of immune-related diseases. While it is clearly important to minimize our exposure to dangerous microbes, is there a way to safely regain interactions with beneficial environmental microbes without risking serious infectious diseases?

LOSING OUR CLOSEST FRIENDS

There are two major ways in which our bodies have contact with microbes on a daily basis: through our resident microbial inhabitants (our microbiota) and the microbes we encounter in passing, from, say, our computer keyboard or by shaking someone's hand. Mounting evidence suggests that decreases in total combined microbial exposure, either from a low-diversity microbiota or an ultraclean environment, can translate into immunological problems. For example, antibiotic use in children (which decreases the diversity of the microbiota) correlates with an increased risk of developing asthma, with each additional course further increasing the risk. But in households with a dog, the risk is reduced. As the hygiene hypothesis would predict, the presence of a dog increases the child's exposure to environmental microbes, thereby mitigating the loss of internal microbes due to antibiotic use.

It is important to note that these studies have not established a causative role for antibiotics and the development of immune-mediated disease, getting back to the chicken-and-egg problem we mentioned earlier. Antibiotic use in humans is related to increased autoimmune problems but it is unknown whether killing your resident microbiota through antibiotic use is the cause of these problems. In the study of human populations, confounding factors make causation very difficult to establish. For example, people taking antibiotics are also typically sicker, have more immune system problems leading to illness, and show a variety of other differences compared to those who rarely take antibiotics. Regardless of causation, there is evidence that the microbiota can protect against autoimmune disease. Germ-free mice, raised in the extreme conditions of no microbial exposure, develop severe airway responses similar to asthma when allergens are present. Mice that are colonized by a full microbiota are protected.

In addition to the collateral damage sustained by the microbiota upon antibiotic use, a second and more specific type of diversity loss has affected members of our microbiota. While oral antibiotic use results in a loss of gut microbes in the short term, over time the microbiota rebuilds, although whether it ever fully recovers is not clear. However, throughout evolution there were species of microbes commonly found in humans, including certain types of bacteria and helminths (a group of parasites such as pinworms or hookworms). Because humans have associated with them for millennia, these species have become known as "old friends." Some of these old friends are capable of causing disease, but over the course of our prolonged friendship, our immune system has come to rely on interactions with these species to function properly. As we modernize, many of these species are being lost, leaving our immune system without the old friends it once interacted with. These companion microbes are still present in the developing world, but as Western culture spreads, increased sanitation, antibiotics, poor diet, and many other factors are leading to their eradication. Although most people consider us better off without helminths, and few people are aware of the important bacteria that we've lost, it is possible that the absence of these old friends may result in an immune system prone to allergies and autoimmunity.

THE IMMUNE SYSTEM BALANCING ACT

An array of immune cells lives within the tissue that lines the intestine, constantly surveying the gut environment and ready to kick into action if necessary. These cells are referred to as the mucosal immune system and their job is to guard against the bad bacteria whose mission is to infiltrate our intestine and cause infection. The mucosal immune system is a subsystem of the immune system that monitors interactions with microbes along body surfaces that are vulnerable to exposure by pathogens. This arm of the immune system protects the tissues within the lungs, nose, eyes, mouth, throat, and gut, all of which have daily interactions with the outside environment. The mucosal immune system is constantly on the lookout for infectious microbes trying to find their way into the body. In the gut its job is twofold: protecting against the occasional ingested pathogen and communicating with and monitoring residents of the microbiota.

The mucosal immune system contains two branches, one that reacts aggressively to a threat (the pro-inflammatory side) and one that dampens the aggressive response once the threat subsides (the anti-inflammatory side). The proper response to gut microbes involves a continual balancing act between these two branches, much in the way that a seesaw is balanced when equal weight is placed on both sides. When the seesaw is perfectly balanced, immune harmony is achieved. The mucosal immune system excludes microbes from breaching the intestinal wall but also keeps the intestinal wall from becoming overly inflamed. Under these circumstances gut microbes and human intestinal tissue live in peace. However, if the pro-inflammatory branch of the seesaw outweighs the anti-inflammatory branch, the unbalanced immune set point can cause an overzealous attack on the resident microbes and spiral into disease. Unfortunately, once the seesaw is off-kilter, it can be difficult to rebalance.

Crohn's disease and ulcerative colitis are two broad categories of inflammatory bowel diseases (IBD) in which patients suffer from inflammation in the distal gastrointestinal tract. Although the causative agents of IBD are still poorly understood, it is clear that both genetic and environmental factors contribute to the onset of these diseases. Numerous genetic mutations have been linked to the development of IBD. Some genetic mutations can result in IBD-like inflammation in laboratory mice, but often on one condition—that the mice have a gut microbiota. In many cases, mice raised in a sterile environment and lacking gut microbes fail to develop disease. The genetics set up the IBD risk like teeing up a golf ball, but it's the microbes that swing the club and send the ball sailing into the rough.

Treatment for IBD usually involves an attempt to rebalance an immune system that too strongly favors the pro-inflammatory side. Immunosuppressant drugs are used to dampen inflammation and antibiotics knock down the resident microbes to minimize the perceived threat. But once an inflammatory response against the resident microbes has begun, it can be difficult to tune down, which is why treatment for IBD can be so difficult. Often surgical removal of the inflamed portions of the intestine is the only solution.

The difficulty in treating IBD illustrates how achieving just the right balance of inflammation is delicate. If there is too little inflammation, microbes can invade the intestinal tissue. Too much can send the immune system into full-blown microbiota-perpetuating inflammation mode. Individuals who are immunocompromised due to chemotherapy treatment or HIV infection offer an example of the hazards of too little monitoring by the immune system. They have a much higher risk of microbial invasion into the intestinal tissue because of their crippled immune system that can't adequately enforce the strict "no microbes allowed" sign outside the intestinal wall. Alternatively, individuals with hyperaggressive immune responses can produce an overly active inflammatory state of interactions with microbes. Some types of immunotherapy treatment for cancer can set up this type of pro-inflammatory scenario by taking away safeguards, or brakes, that naturally arrest an inappropriate immune response. The hope with this type of treatment is that by favoring the pro-inflammatory side, the aggressively tuned immune system will attack the cancer cells. The danger, however, is that friendly bacteria within the intestine can also become targets, leading to IBD-like disease. These clinical situations demonstrate the precarious nature of immune homeostasis and monitoring of our gut residents.

Unfortunately, it's not just immunocompromised people or those on immunotherapy who have to worry about maintaining a healthy immune system balance. It appears that the Western lifestyle has also disrupted the seesaw, jeopardizing the delicate balancing act that keeps both the pro- and anti-inflammatory branches of our immune system peacefully coexisting with our microbes. More and more evidence is building that gut microbes aren't just bystanders in the immune system balancing act. The microbiota plays an active role in setting the tone for how the immune system responds to both gut microbes as well as foreign pathogenic microbes.

MICROBES AS AN EXTENSION OF THE MUCOSAL IMMUNE SYSTEM

The gut lining is protected by a gooey shield of mucus, a physical barrier that prevents gut microbes from wandering too close to human tissue. This layer of slime not only keeps the microbiota at a safe distance, but also serves as a rich source of carbohydrates that some bacteria within the microbiota can eat. By producing this carbohydrate-rich secretion, the gut provides nourishment to help sustain certain beneficial members of this community. These microbes in turn help defend the gut from invasion by bacterial pathogens and balance the immune system.

The pathogenic *E. coli* from the undercooked hamburger you ate arrives in your digestive tract hoping for a quick and easy entrance into your intestinal wall. But as this pathogen attempts to penetrate your body's internal surfaces, before it even tries to bushwhack its way through the mucus layer it must first contend with a gauntlet of resident microbes.

The gut bacteria serve as the first line of defense against these invaders, providing both a physical and biochemical hindrance to pathogenic forced entry. The microbiota is like a mercenary in the eyes of the immune system, paid (in slimy mucus) for helping to exclude bad germs but not trustworthy enough to go completely unmonitored.

The microbiota does more than serve as an extra barrier to pathogens; it also tunes the magnitude and duration of the immune system response, much as a puppeteer controls the strings of his puppet. For example, when invaders are present, if the immune system mounts a slow or apathetic response, the pathogen gains an advantage. Alternatively, if the immune system is overzealous, excessive inflammation and unnecessary tissue damage or autoimmunity may result. The resident microbes are in many ways holding the strings that control your immune system, helping the immune system determine the strength and pace of its response. The microbiota fine-tunes the immune

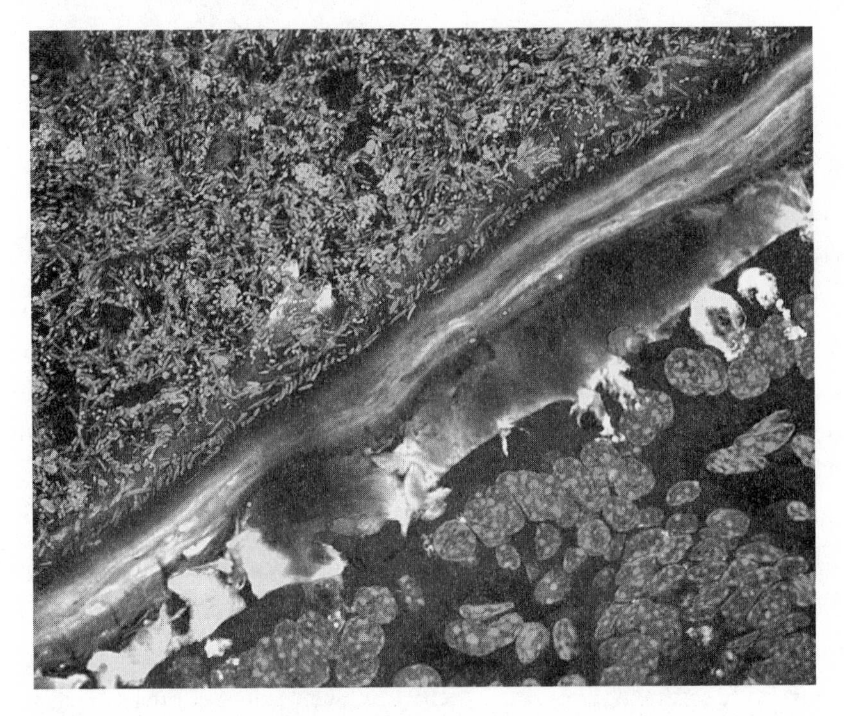

A fluorescence microscopy image of the microbiota within the intestine. Resident microbes are the rods covering the top left corner. Mucus is the thick striped layer bisecting the image. Intestinal cells are located in the bottom right corner.
© Kristen Earle and Justin Sonnenburg

system as it undergoes developmental changes throughout life. Immune system development is most rapid and noticeable in the first few years of life, as a fetus protected by the womb makes the transition to an infant and then a toddler confronted with trillions of microbes. It turns out that the exposure to microbes, the very entities that the immune system must monitor, is critical, especially early in life, for proper immune system development.

From Chapter 2, we know that mice with no microbiota have a thin and patchy mucus layer coating their intestine compared to mice with a conventional microbiota. Remove the microbes, and this critical element of the mucosal immune system does not form properly. Beyond the mucus layer, there are other profound differences in the appearance, composition, and function of the mucosal immune system in these microbiota-free mice. They have almost no intestine-resident immune cells, which are needed to respond to microbial advances. Some of the immune system deficiencies in germ-free mice can be corrected if the mice are colonized by a full microbiota. However, in some cases the deficiencies are not correctable. If microbial exposure happens too late in life, a critical early window of time during development is missed and the immune system is locked into an underdeveloped state. Think of forgetting an ingredient while following a recipe. In the case of soup, forgetting to add the salt can be easily corrected as a final added step, and the soup will taste the same as if you hadn't forgotten. On the other hand, if upon pulling a cake out of the oven you realize that you forgot to add the baking powder when you mixed the batter, there is no amount of baking powder you can add at that point that will fix the flat cake!

Although going through the beginning of life without a microbiota would never occur in humans, it is possible to go through the first weeks of life with fewer microbes due to antibiotic treatments and life in our overly sanitized environment. Microbes that we interact with during the critical early time of life may dictate aspects of our immune system's maturation that are irreversible. The lesson here is that it is possible that raising children in an overly hygienic environment could have a long-lasting detrimental impact on the development of their immune systems.

BALANCING THE IMMUNE SYSTEM WITH MICROBES

The responsiveness of the immune system to gut resident microbes means that promoting specific members of the microbiota has the potential to optimize immune function. Is it possible to determine which microbes are best for promoting a healthy immune system and create the ultimate "immune boosting" probiotic supplement? A pill of beneficial bacteria that would create the perfectly balanced immune system, one that quickly fights off infection yet leaves pollen or peanuts alone, is just what we need, right?

Unfortunately, the complexity of the mucosal immune system makes this possibility seem more like science fiction than science. Immune responses are commonly dictated by B cells and T cells kicking into high gear in a pro-inflammatory surge, bringing redness, swelling, heat, and pus. But the flip side of this response is the attenuation of the redness, swelling, heat, and pus after the response is mounted. This job is accomplished by a type of immune system cell called the regulatory T cell or T-reg. A paucity of T-regs can lead to an overzealous immune response, which in turn can progress to autoimmunity, inflammatory bowel disease, and even cancer. Some hypothesize that a deficiency in T-regs is a hallmark of many Westerners and is the basis of many Western diseases. The recruitment of additional T-regs, if such were possible, could lead to new treatments or preventative strategies for many inflammatory diseases.

Kenya Honda's research group at the RIKEN Center for Integrative Medical Sciences in Japan has discovered that members of the microbiota are responsible for populating gut tissue with T-regs. Honda is in the league of scientists contending that our modern microbiota has deteriorated due to a number of factors, including antibiotics and poor diet, "which renders the host more prone to induce autoimmunity and allergy." He notes that the skyrocketing rates of patients with inflammatory bowel diseases, allergies, and multiple sclerosis has markedly increased for several decades and is still increasing in his home country of Japan.

Honda and his team discovered that members of the Firmicutes (one of the two major phyla of bacteria in the gut microbiota) are capable of recruiting T-regs to the gut of laboratory mice. This abundance of T-regs dampens inflammatory responses and makes the mice less likely to develop colitis, autoimmune diseases, and allergies. This mixture, or cocktail, of intestinal bacteria can modulate the mammalian immune system in ways that no known drugs can. The question is, since everyone has a different microbiota, would the same exact cocktail of bacteria work in each person? "I am quite sure that differences in the microbiota between people will be a relevant factor," says Honda. The chances that a specific cocktail of bacteria would result in the same anti-inflammatory effect in everyone seem slim. But perhaps the species of microbes is not as important as the molecules they produce.

As microbes consume food within the intestine, they produce by-products that amount to waste—bacterial feces (yes, our intestine is a microbial toilet). While it's not exactly pleasant to think about, many of these bacterial waste products don't appear to be as harmful as you might expect. In fact, some of their waste products are health promoting. One of the most abundant microbiota waste products is short-chain fatty acids, or SCFAs (more on these special molecules in coming chapters). These molecules help the intestine accumulate T-reg cells. For the gut microbiota it may matter less "who is there" than "what they are doing." Many different types of bacteria are capable of producing SCFAs, therefore promoting SCFA production by the bacteria already present in the gut microbiota could promote T-regs and dampen inflammation. Although these studies are preliminary, they provide hints about how we should think about manipulating our gut microbes to improve our intestinal health. Magic combinations of microbes that confer great health benefits may not be commercially available for some time. But nudging the microbiota to generate more SCFAs and other important chemical messengers that balance the immune system could help to prevent or ameliorate disease.

ASSESSING WHETHER A MICROBIAL TENANT IS BAD AND THE COSTS OF EVICTION

The immune system has the enormous task of ridding our bodies of a collection of bad guys. But in many ways getting rid of bad guys is the easy part. The immune system has evolved an arsenal of both strategic weapons (highly targeted antibodies) and weapons of mass destruction (like fever and diarrhea) that are highly effective at neutralizing bad guys. The more difficult job of the immune system is to determine who is a bad guy and who is not. If the immune system gets it wrong, a potentially dangerous infection could be ignored or, as is the case with multiple sclerosis, a completely normal and critical set of cells can be attacked. People can also have problems properly categorizing good and bad bacteria, especially since many bacteria fall into a "gray" zone. These gray zone bacteria may be harmful in some situations or in certain individuals, but may be beneficial in other instances.

Martin Blaser is a professor at New York University and a leader in the study of the health impact of the stomach-residing *Helicobacter pylori*, also known as *H. pylori*. This bacterium can cause stomach ulcers and sometimes stomach cancer. Clearly a bad guy, right? The medical community thought so and has targeted this "bad" microbe for elimination using antibiotics.

"It's test and treat," in Blaser's words. "As a doctor, if you find *H. pylori*, you get rid of it. Yet, if you look at the evidence, there are very few people that should be treated for Helicobacter." While *H. pylori* can be problematic for certain individuals, many people are unaware that they harbor this bacterium and never suffer any ill effects from it. In fact, evidence is mounting that having *H. pylori* may even be beneficial.

H. pylori is usually acquired from a person's parents, so if prospective parents are treated and lose *H. pylori*, they can no longer pass it on to their children. This scenario is being played out in the Western world. Over just a few decades, ever since *H. pylori* got its "bad bacterium" label, the microbe is gradually being eradicated. With each generation, fewer Western children have *H. pylori* in their stomachs. On the surface this may seem like great news—these children won't develop *H. pylori*–induced ulcers or stomach cancer. But there is a potential giant downside to *H. pylori*'s extinction within our population. As studies by Blaser and others have shown, children who are not colonized by *H. pylori* are at increased risk for developing asthma and allergies. While the absence of *H. pylori* may protect a small minority from stomach problems later in life—few people harboring *H. pylori* will develop stomach ulcers or cancer—its absence at the beginning of life may put children at risk for a lifetime of health issues. It is likely that this bacterium, which has coevolved with humans for tens of thousands of years, helps regulate our immune system to an optimal set point. By eliminating this microbial "teacher," the immune system loses some of its ability to discriminate between an appropriate target, such as a flu virus, and an inappropriate one, like pollen. Loss of *H. pylori* may only be the tip of the iceberg. As we learn more about what types of bacteria and other microorganisms inhabited the gastrointestinal tract of our ancient ancestors, it is clear that modernization has led to the eradication of several previously prominent members.

The disappearance of *H. pylori* demonstrates two important points. First, even a single type of bacterium can be instructive to the immune system. Before targeting bacteria for elimination, especially those that have been associated with humans for thousands of years, the potential damage to our immune system needs to be considered. Second, some of our associated bacteria exhibit Jekyll and Hyde personalities. We don't yet completely understand the factors that can turn a seemingly friendly bacterium into a pathogen. Attaching labels such as symbiotic or pathogenic to bacteria is an oversimplification that disregards the ability of a microbe to change its personality depending upon the situation.

How will this emerging understanding of the nuances and complexities of human being–microbe interaction impact human health? "Doctors of the future will be giving *H. pylori* back to children and then eradicating it at an older age," says Blaser. Eliminating *H. pylori* after reproductive age would ensure that this microbe and its benefits are passed on to future generations.

GETTING THE IMMUNE SYSTEM DIALED JUST RIGHT

The set point of the immune system follows the Goldilocks principle. If it is dialed to a point of being too "hot" (i.e., overly reactive), autoimmune disease could follow. If it is set too "cold," legitimate infections could be ignored or allowed to fester. Ideally, the immune system is set just right: dangerous infections are being addressed and our own cells and friendly microbes are tolerated.

H. pylori demonstrates the power that our gut microbes can have in governing our immune system. But it is just one of our many microbes that are capable of shaping immunological parameters throughout our bodies. Microbes that were traditionally a part of the human being–microbial superorganism have since been lost in subsets of the modern human population. Pieces of our biology—gone. Humans are dependent on their resident microbes. When these critical associations are broken, for example when species of bacteria are eradicated, deficiencies and diseases can result. The specific details, or context, of each association are of supreme importance when determining whether a microbe is ultimately beneficial or not to the life of the host.

Knowing the importance of the microbiota–immune system connection, are there choices that we can make to positively impact our microbiota and immune health? Compounding the complexity of navigating the intricacies of the human immune system is assessing the equally complex nature of our highly individualized resident microbial communities. Excitement within the scientific community over the realization that the microbiota can modulate the immune system is steadily growing. But for safety reasons, researchers looking for therapies that exploit this knowledge must proceed with caution. The same applies to scientists dispensing health advice. However, we feel that there is sufficient evidence to take actions that are safe and that positively impact our microbiota and health. (Of course it is always prudent to consult your physician regarding any suggestions to ensure they are compatible with your specific medical history.)

Parents hoping to raise children free of allergies and asthma often ask us: Should I have my children wash their hands before eating? Should we own a dog? Do my children play in the dirt frequently enough? Clearly no single correct answer exists for any of these questions. The answers should result from a cost-benefit analysis, in which the pros and cons are weighed for each situation.

Here is how we decide what to do when confronted with these situations.

Our personal approach to hand-washing is one example of how to be proactive about an emerging body of data, even in the absence of a definitive scientific study. We often do not have our children wash their hands before eating if they have just been playing in our yard, petting our dog, or gardening. However, after visiting a shopping center, hospital, petting zoo, or other area that is more likely to harbor pathogens from other humans or livestock, washing hands is mandatory. We also increase the frequency of washing during cold and flu season or if we have potentially come into contact with chemical residues (e.g., pesticides). We are aware of the risks of acquiring pathogenic microbes, and with the existence of antibiotic-resistant superbugs the stakes are high and decisions are not to be taken lightly.

However, the epidemic of autoimmune disease in modern societies suggests that sanitization and hyperhygienic practices are not a panacea.

To the pet question. Owning a pet is a huge responsibility and one that should not be taken on for the sole purpose of microbial contact. There are other, less labor-intensive ways to accomplish this. However, we view the extra microbial exposure that comes with owning a pet as an added bonus to all the other advantages a pet can provide, from companionship to an excuse to go for a walk every day (both of which provide their own health benefit). People who own dogs have bacteria on their skin that is similar to that of their dog, but unlike other dogs'. It's no surprise that physical contact with a dog transfers microbes, probably in both directions. The transfer of bacteria from dogs results in a more diverse skin microbiota for their owner. It is likely that dogs acquire microbes on their fur from parts of the environment that humans are less exposed to, like the side of a fire hydrant. The increase in microbial diversity owning a pet provides may partly explain the observation that children who grow up around pets are less likely to suffer from allergies or asthma.

People not inclined to owning a pet needn't worry. Dirt is another way to increase your exposure to environmental microbes. Some scientists estimate that the bacterial diversity in a typical soil sample is about three times greater than that found in our gut. That stuff we're constantly wiping off our shoes and telling our kids to clean off their hands is a veritable wilderness filled with microorganisms. Unfortunately, our fear of encountering a lion has caused us to eliminate even the most gentle deer. Scientific evidence is mounting that exposure to environmental microbes like those found in soil may help protect us from autoimmune diseases.

But before you send your kids running into the backyard to make a mud pie or remove your doormat to allow more microbes into your home, understand that there is a cost-benefit analysis that needs to be considered. Unlike the ground-level environment of our ancient ancestors, who dug for edible tubers and lived in dirt-floored homes, the dirt of the modern world is often laden with an assortment of man-made chemicals such as fertilizers, weed killers, and insecticides. Ingestion of these compounds can potentially negate any positive benefit soil microbes may have. So, if your yard is not chemically treated, then digging around in the soil and not washing up immediately may offer a benefit of added microbes without the risk of chemical exposure. If you bring your children to a playground with an expanse of grass that doesn't appear to have a single weed in it, then washing hands afterward is probably the safest course of action. Gardening at home, even if it's just in a container, can be a great way to expose yourself to healthy, compost-infused, bacteria-containing soil.

Current knowledge indicates that most of us living in the modern world could benefit our immune system by increasing our microbial exposure. How you accomplish this should be in a way that is safe, comfortable to accomplish, and compatible with your lifestyle.

CHAPTER 4

The Transients

A CRY FOR HELP

Recently a close friend contacted us for some much-needed advice. Rick is generally healthy but occasionally experiences digestive issues ranging from constipation to bloating. His doctor had recommended, in passing, that he try probiotics as a way to help. The hope implicit in the recommendation is that by supplementing the gut with probiotics, these beneficial bacteria can restore better digestive health; and probiotics may even help the immune system function in a more balanced manner. One quick trip to the drugstore later, Rick was understandably overwhelmed by the variety of probiotics that were available. He came to us with a litany of questions: Which probiotics are best for someone like me? How often should I take probiotics? Will probiotics even help my digestive system? Should I take them in supplement form or get them from food?

Probiotic means "for life" and was officially defined by

the World Health Organization as "live micro-organisms which, when administered in adequate amounts, confer a health benefit on the host." This definition, however, leaves a gray area for collections of microbes that offer potential health benefits (like those found in fermented foods) but fail to qualify under the formal definition simply because they haven't been studied. For this reason, and for the purposes of this book, we will use the term "probiotic" to refer to consumable bacteria that may provide a health benefit or are marketed as providing a health benefit.

Probiotic bacteria, unlike the long-term bacterial residents that make up our microbiota, are transient visitors to the gut. But their fleeting transit through the gut does not mean that their impact is unfelt by us and the microbes we house. There is growing evidence that probiotic bacteria can help reduce our chance of catching certain infections and, when we do, help us recover from them faster.

Probiotics offer a method distinct from diet to tune the microbiota and together with diet can impact health in a beneficial manner. As they drift through the digestive tract, probiotics communicate with the resident microbes and intestinal cells. The immune system profits from these interactions, as people who consume probiotics are better able to fight off colds, flu, and diarrheal illness. While there is much conflicting information about probiotics, consuming bacteria is a practice as old as humanity itself. The gut has not only evolved to deal with the constant passage of ingested bacteria, but has also learned how to benefit from these daily guests.

THE BIRTH OF FERMENTATION

What is your most valuable kitchen appliance? Think not in terms of how much it cost, but rather of the one you feel you can't live without. Many people would choose the refrigerator. When we lived in St. Louis our house happened to be on a borderline for the power grid. Frequently, during a storm, the power would go out, but only on one side of the street. Sometimes, depending on the severity of the storm, there would be no power for days, leaving half of the neighborhood in the dark, while the other half lived a typical kilowatt-consuming lifestyle. But within twenty-four hours of the outage our street would be littered with extension cords piping electrons from a house with electricity to the refrigerator of the less lucky neighbor across the road, a visual reminder with each storm of just how precious a functioning refrigerator is.

How did people keep food fresh before the modern refrigerator, before the wooden armoire-like icebox, even before the underground pits filled with snow or ice used by the ancient Greeks, Romans, and Chinese? How did our prehistoric ancestors, especially those living in tropical climates with no access to ice or snow, keep food from spoiling? The answer is that often they didn't, but instead learned how to control spoilage in a way that allowed rotten food to remain edible.

Fermentation is the process by which microbes consume sugars and produce acid, alcohol, and gases. Some of the more well-known products of fermentation are wine and beer, where the sugar in fruit juice or grain is transformed into alcohol by yeast. Today we enjoy the inebriating effects of fermentation, just as our ancestors did, but perhaps more important to them, alcohol functioned as a preservative, increasing the time during which the beverage could be stored. Similarly, food can be fermented by bacteria to extend its shelf life. For example, some types of fermented cheeses stored at room temperature can still be edible many years later.

The discovery of fermentation was likely an accident. Perhaps an especially large amount of food was gathered, too much to be eaten all at once. Since we know that every calorie was precious, there would be a huge incentive for our ancestors to not waste food that was starting to go bad. As they began to understand that some rotting food was still safe to eat, they learned to use fermentation as a way to make the food supply a little more stable.

Refrigerators are used in much the same way today, keeping food edible longer. The one big drawback is that our modern method of preservation largely excludes microbes from our diet. We know that the ability to organize into social groups and divide labor allowed society to advance beyond its hunter-gatherer roots. Perhaps our ancestors' ability to control fermentation removed an additional labor burden, freeing them to advance in other areas.

The earliest record of fermented food consumption is more than eight thousand years old, and at least one type of fermented food is a part of virtually every culture's history. Through fermentation, living bacteria start the process of digestion for us. One of the most familiar fermented food products is yogurt. To make yogurt, certain types of bacteria are added to milk, a rich source of the sugar lactose. The bacteria ferment the lactose, converting it into lactic acid, which provides the characteristic tartness of yogurt. You can think of that container of yogurt in the fridge as an external digestive tract, predigesting the lactose before it ever enters your mouth. This means that yogurt can be eaten by some who are lactose intolerant, but for those who can digest lactose, there is some forfeiture of calories to the bacteria. In the past, these sacrificial calories would have been a small price to pay in return for the longer shelf life it provided. In the cheap and easy calorie-rich modern world, losing calories to bacteria is not such a sacrifice. In fact, when microbes ferment food they deplete it of its simple sugars (like lactose in the case of yogurt). Too much simple sugar can cause blood sugar to spike, which can lead to health problems including type 2 diabetes. Bacteria, through fermentation, reduce the simple sugar content of food, making it a little healthier. The microbes in fermented food provide two health-promoting functions: reducing the sugar content of food, and interacting with the gut and microbiota. Observations dating back over a century suggest that people who consume a lot of fermented foods reap the benefits.

PRESERVING THE GUT

Élie Metchnikoff, the late-nineteenth-century Russian-born scientist, was especially interested in microbes and how they interact with the immune system. Using a microscope he observed how a certain type of immune cell in our blood responded to an invader, engulfing it the way Pac-Man consumes pellets. Joining the Greek words *phages*, meaning "to eat," and *cite*, meaning "cell," he called these microbe-eating cells phagocytes. With the discovery of these Pac-Man-like cells, Metchnikoff had revealed a major strategy employed by the immune system to rid itself of pathogenic microbes and was rewarded with a Nobel Prize.

Near the end of his career, Metchnikoff became keenly interested in how humans age and die. In 1908 he recorded his scientific findings and ideas in a book titled *The Prolongation of Life: Optimistic Studies*. He proposed that aging and death are the result of a buildup of toxic waste in our gut, waste generated by the bacteria living there. Metchnikoff viewed the large intestine as a mostly useless organ that evolved as a storage container for fecal waste. (Those of us who have spent most of our adult life studying the microbiota try to chalk up Metchnikoff's view as just slightly simplistic and not insulting.) He reasoned that in order to effectively hunt prey, man needed a way to keep from continually defecating. "A carnivorous mammal which, in the process of hunting his prey, had to stop from time to time, would be inferior to one which could pursue its course without pausing," he explained. But to Metchnikoff, the storage of fecal waste within the large intestine came at a price. While some of the gut-residing bacteria "are inoffensive, others are known to have pernicious properties." He was convinced that these pernicious bacteria were the reason that humans couldn't enjoy a longer life span. Noting that adding acids to food could prevent "putrefaction," or rotting, Metchnikoff reasoned that humans could minimize their internal putrefaction by the aid of acids as well, specifically lactic acid.

He became convinced that consuming bacteria that produced lactic acid (like the bacteria found in yogurt) would "preserve" the gut in the same way that these bacteria can preserve milk. Although his mechanistic explanation for why fermented foods are healthy needed some revision, Metchnikoff's observations, including the long life span of Bulgarian peasants who drank fermented dairy products daily, began to reshape contemporary thinking about microbes. His recommendation, more than one hundred years ago, to consume more bacteria, specifically lactic-acid-producing bacteria, awakened the scientific community to the idea that the consumption of fermented dairy products could increase life span. In his wisdom he wrote, "A reader . . . may be surprised by my recommendation to absorb large quantities of microbes, as the general belief is that microbes are all harmful. This belief, however, is erroneous."

As our understanding of how probiotic bacteria function within the gut has matured, it is clear that Metchnikoff's view that their benefit is derived from their ability to acidify the gut is not the complete story. These bacteria, although they account for such a small amount of the total bacteria living in the gut at any one time, are capable of impacting our biology in ways that seem out of sync with their abundance. They can even exert their effects beyond the gut, sending signals to the farthest corners of the body, including the brain.

GUT TOURISTS: PASSING THROUGH AND LEAVING THEIR MARK

One common misconception about probiotics is that these living bacteria take up permanent residence in our gut. Probiotic bacteria are typically only transient members of our microbiota, passing into the gut upon consumption and then exiting. In the case of the *Lactobacillus* that are found in fermented dairy products, they are most at home in environments that contain lactose, like milk. Even breastfed babies are typically not colonized by lactic acid bacteria, like those that ferment dairy products, because lactose from mother's milk is digested and absorbed by the infant and not available to microbes that live in the colon.

So while many probiotic bacteria can survive in our gut, most are not well suited for this environment. They are not equipped to consume the exotic foods found in our gut—like the dinner we ate or the mucus layer coating our intestine. Therefore these bacteria are only temporary residents that pass through our digestive tract. A reason why proponents of probiotics recommend their regular consumption is to ensure a steady stream passing through. Probiotic bacteria are similar to tourists visiting a foreign land—our gut—from their native land—the yogurt or other fermented food that they grew in.

The fact that these bacteria don't dwell within the gut for long and are not very numerous relative to the resident bacteria does not mean they are just inert entities. There is evidence that the presence of probiotic bacteria passing through us reinvigorates our body's defenses against invading pathogens. In a way probiotics can serve as "dummies" that allow the immune system to fine-tune its response to more dangerous microbes.

The cells that line our intestinal wall sit side by side, like tiles. In between these cells is a network of proteins that serve as the grout. The grouted, tiled wall is the barrier that keeps the microbiota and particles of digesting food from crossing into our tissue and bloodstream. Ideally, bacteria stay within the boundaries defined by the tiled wall, that is, inside the tube. Studies suggest that probiotics can help reinforce the gut barrier by nudging intestinal cells to produce more protein "grout." In addition to fortifying the so-called tiled wall, probiotics can also promote the secretion of mucus—the slimy shield that protects us from unwanted invaders—that lies on top of this wall.

If fortifying the intestinal wall and increasing the mucus layer weren't enough, probiotic bacteria also can coax our intestinal cells to release molecules known as defensins, one of the body's chemical warfare agents against invading bacteria, viruses, and fungi. Which specific probiotic strains are involved in these methods of safeguarding our gut and how they accomplish this task requires more research. As a better understanding of the positive responses within the gut and probiotics emerges, maybe in lieu of tourists it would be more accurate to think of them as UN peacekeepers, helping to enforce borders and deter aggressive forces.

Probiotic bacteria, through their ability to reinforce the intestinal border and prime the immune system, should be effective allies in the fight against gastrointestinal infections. To test this notion directly, a group of researchers from Georgetown University Medical Center looked at whether consuming probiotics could protect young children from gastrointestinal infections. They recruited 638 children between the ages of three and six years old who were attending a day care center in the Washington, DC, area. Half of the children were randomly assigned to consume a fermented dairy drink containing probiotic bacteria and the other half drank a placebo containing no bacteria daily, for ninety days. Parents filled out weekly questionnaires that addressed the health of the child, asking if they had missed school due to illness; if they had experienced vomiting, constipation, stomach pain, or fever; or if they had been prescribed antibiotics. Children who had consumed the probiotic drink were 24 percent less likely to have suffered from gastrointestinal infections compared to the children not receiving probiotics. The probiotic-consuming children were also less likely to have used antibiotics during the course of the three-month study.

This study is not alone in documenting the ability of probiotics to protect against gastrointestinal infections. A number of studies have shown that probiotics in general (not one specific strain or product) can have a positive impact on people suffering from infectious diarrhea, reducing both its severity and duration. These beneficial microbes, either by reinforcing our intestinal barrier or by directly or indirectly killing pathogens (or perhaps by an as yet undiscovered mechanism), are able to fend off and in some cases lessen the persistence of the infection. Probiotic bacteria, while not permanent residents of our microbiota, can be allies in our skirmishes with pathogens in the gut.

NOT JUST A GUT EFFECT

It makes sense on an intuitive level that consuming probiotic bacteria can impact health within the gut. They are in close proximity to the intestine and the microbiota as they travel through. But, similar to the resident gut microbes, probiotic bacteria, as they drift through the digestive tract, appear to impact our biology beyond the gut and serve to promote health throughout the body.

When studying the children at the Washington, DC, day care center, researchers found, somewhat unexpectedly, that not only did the probiotic-consuming children have lower rates of gastrointestinal infections, they also had fewer upper respiratory tract infections as well. Other trials encompassing thousands of people have also found fewer acute upper respiratory infections and less antibiotic use among probiotic consumers of all ages. These findings are building support for a view of probiotic bacteria that takes into account their ability to tap into the functioning of our immune system, not just within the local environment of the gut, but on a global scale.

Several studies have shown that probiotic consumption among healthy human volunteers coincides with changes in immune system function to aid in the fight against infections. The immune system seems to be constantly performing a census of the microbial life in the gut. When probiotic bacteria are present, the immune system adopts a ready state, the equivalent of "on your mark." When an infection appears, even if that infection is in the upper respiratory tract, the immune system is poised to "get set" and then "go."

Wait a minute—if this is the case, why aren't doctors everywhere encouraging probiotic consumption? Here's the catch: of the thousands of published studies performed on probiotics, most focus on relatively small groups of people, and their specific findings have not been replicated in other studies. In addition, few studies in humans are able to ascribe specific beneficial effects to specific strains of probiotic bacteria. This failure to reveal the mechanisms, or specific molecular interactions and genes, that are involved in, for example, a specific immune system effect, increases skepticism.

Why do probiotic effects seem so erratic? When an individual ingests probiotic bacteria, those bacteria interact with the microbes living within that person's microbiota. Because each person's microbiota is unique, probiotic A in Person 1 may behave differently than it would in Person 2. Maybe Person 2 needs to consume probiotic B or ten times more probiotic A to see the same effect observed in Person 1. Since the microbiota can fluctuate from day to day, even within an individual, a probiotic's effect may also vary over time. Presently, our understanding of the microbiota is not complete enough to predict what specific effects a particular probiotic could have on an individual's microbiota. For this reason, we feel that fermented foods, which contain a diverse collection of microorganisms, offer the best chance of encountering a microbe that will have a positive effect.

The most common type of probiotic-containing foods in the United States is fermented dairy products like yogurt and cultured sour cream (although sour cream can also be produced without bacteria, so not all sour cream contains living microbes). Kefir, a less common probiotic dairy beverage, is fermented using a diverse collection of up to a hundred different species of bacteria and yeast and, like yogurt, contains billions of living microbes per serving. Its vast assortment of microorganisms makes it a favorite in our household, especially during cold and flu season. The diverse set of microbes in kefir maximizes the chances that, within our family of four, each person's microbiota will respond to at least one of the microbes within this drinkable probiotic zoo.

In the West, other familiar fermented foods include sauerkraut (fermented cabbage), pickles (fermented cucumbers or other vegetables), and more recently kombucha, a popular fermented, sweetened tea. In addition to these more readily available options, people around the globe have figured out how to ferment just about everything, including beans, fruits, vegetables, grains, and even meat and fish. Hakarl, a traditional dish in Iceland, is shark meat that has fermented for up to three months in a sand- and gravel-filled hole dug into the side of a hill. (The angle allows for the draining of juices.) Although we cannot personally attest to the palatability of this dish, we imagine it must be an acquired taste!

PROBIOTICS: WHAT QUALIFIES?

Recently, a huge industry has sprouted to deliver more probiotic supplements and probiotic-containing fermented food products. The companies manufacturing these products are hoping to convince us of the significant health benefits that come with consuming probiotic microbes in food or in purified supplement form.

The Web is filled with sites lauding the benefits of consuming bacteria and selling probiotic supplements to promote gut health. Often these sites are shrouded in unfamiliar terms like synbiotic, functional food, and nutriceutical, words that may instill hope, intimidate, confuse, or perhaps do all three. According to many of these websites, we should be taking these supplements daily and in vast quantities. If you are healthy, these products claim to maintain health or prevent disease. If your gut is unwell, well, here is the solution. With names like Ultimate Flora Super Critical, Primal Defense, and Healthy Trinity these supplements scream, "If you want to be healthy, you need me!"

Adding to the cloud of confusion surrounding these products is the fact that there is little consensus within the medical community regarding who can really benefit from probiotics. However, just over the past few years the scientific evidence has solidified for the use of probiotics in several clinical situations. Mary Ellen Sanders, PhD, an independent consultant in the field of probiotics who serves as executive director of the International Scientific Association for Probiotics and Prebiotics (ISAPP), agrees that there is now compelling data to support the use of probiotics for a number of conditions such as necrotizing enterocolitis for premature infants, antibiotic-associated diarrhea, acute diarrheal illness, and even the common cold.

Unfortunately, because reproducible scientific conclusions regarding the use of specific probiotic strains to treat specific illnesses have been lacking, much of the medical community operates under the assumption that probiotics are unlikely to hurt, and possibly may help. This may be a reasonable approach considering the excellent safety profile and many promising preliminary studies.

Dr. Purna Kashyap is the associate director of the Microbiome Program within the Center for Individualized Medicine at the Mayo Clinic in Rochester, Minnesota. Although officially a gastroenterologist, he spent two years training in our lab at Stanford to gain expertise in how the microbiota influences gastrointestinal health. His practice at Mayo focuses on functional gastrointestinal disorders such as irritable bowel syndrome and motility disorders. With regard to probiotics his approach is more passive than active: "If asked by a patient, I don't discourage use but don't suggest it as a first line treatment."

Dr. Kashyap's tentative approach to probiotics is not uncommon among physicians. Many feel wary about recommending a course of action that has not been validated in rigorous and reproducible clinical studies with outcomes that are measurable as opposed to reports of overall "feeling better." Despite his skepticism, Dr. Kashyap regularly drinks buttermilk, a cultured dairy drink that reminds him of the yogurt his mother would make for him daily when he was growing up in India.

WHAT'S IN A NAME?

Most consumers have a positive association with the word "probiotic," giving companies an incentive to use it to market their products even when its use is not justified. According to the ISAPP consumer guide for probiotics, "Just because it says 'probiotic' doesn't mean it is a probiotic. Some products labeled 'probiotic' do not contain strains shown to be effective or may not deliver adequate levels of live probiotic through the end of shelf life of the product." So while probiotics offer much promise, there are many reasons for consumers to be skeptical when a product features the label "probiotic."

There are a number of different types of bacteria marketed as probiotics. But before we delve into the specifics, it's worth taking a few moments to discuss how bacteria are named, since these names may contain information about the properties of the bacterium. Consumers should also be aware that names given to bacteria by companies can be used as marketing tools.

Bacteria are referred to scientifically using two names, a genus name and a species designation. *Bifidobacterium* and *Lactobacillus* are two genera (plural of genus) of the most common types of commercially available probiotics. You can think of a genus as the bacteria's last, or family, name even though it appears first. All bacteria that fall within a particular genus are closely related. The species name is equivalent to a person's first name (although it appears last in the bacterial name) and specifies a particular member of that genus. *Bifidobacterium longum* and *Bifidobacterium animalis* are two different species of bacteria from the same genus. They are more similar to each other than *Bifidobacterium longum* is to *Lactobacillus acidophilus.* Additionally, bacteria of the same genus and species can have distinct strain designations. This denotes smaller variations that can exist within a species. For example, we are all *Homo sapiens* but still retain individual characteristics that distinguish each of us from all others. For bacteria, a particular strain is often denoted as a set of letters and numbers that follow the genus and species name, for example *Bifidobacterium animalis* DN-173-010. Particular strains of bacteria can be proprietary and are often given trade names by the company that owns them. Often these trade names are chosen to evoke associations between the bacteria and digestive health. For example, prominently displayed on containers of Activia yogurt is the name of Dannon's probiotic bacteria: *Bifidus regularis,* the trade name given to their strain of *Bifidobacterium animalis.* This same strain of bacteria has a different trade name depending on the market in which it is available. In the UK it is called *Bifidus digestivum.*

If you think there are regulations in place to keep probiotic companies from misleading consumers, you'd only be partially right. The term "probiotics" is used by a large and varied group of products containing live bacteria, and the FDA has developed a complex framework for regulating these products, depending upon their intended use. Since most probiotic products are not marketed specifically to treat disease, they are not categorized as drugs and are not required to undergo the testing and regulation that govern drug approval. By avoiding a drug categorization, probiotics can bypass scrutiny by the FDA, whose primary concern with these products is safety rather than effectiveness. But because probiotics are not drugs, the FDA prohibits claims on the labels that would indicate they treat disease.

With so much money (and potential health benefits) on the line, you would think that companies selling probiotics would have gone to the trouble of identifying the most beneficial types of bacteria. The reality is that few probiotics found either in food or supplement form have gone through any sort of rigorous selection process (although the companies selling these products may disagree). While a few probiotics have been selected based on special properties, the vast majority of strains have been chosen somewhat arbitrarily from what naturally occurs in fermented foods.

Let's consider the three main avenues by which most people encounter probiotics: (i) foods that are fermented, such as yogurt; (ii) foods to which live microbes are added but that do not perform fermentation reactions, such as probiotic-containing granola bars; and (iii) microbes that are in supplement form. In all cases the specific microbes used have either had a long-standing history of being consumed without ill effect, or have achieved a special status referred to as "generally regarded as safe," or GRAS. To attain GRAS status a consensus of qualified experts must agree that the product is safe to consume, but because of its budget issues, the FDA has essentially made GRAS registration of probiotics a voluntary program.

Let's say I am planning to start my own probiotic supplement company, which I've named Immune Boosters. My first probiotic supplement will contain the bacterium *Lactobacillus casei*, a common species found in yogurt, so I know it's safe and won't raise any red flags with the FDA. I have my own proprietary strain that I will market under the trade name *Lactobacillus ProHealthy*. Before putting bottles of Immune Boosters' *Lactobacillus ProHealthy* in drugstores across the country, I would need to notify the FDA of its ingredients and safety information. Ninety days after this notice is sent, *Lactobacillus ProHealthy* can be made available to consumers—no FDA approval necessary. With the responsibility of the safety of supplements delegated to the companies that sell those supplements, it is not surprising that such relaxed oversight has led to shelves filled with suspect products. In many cases the identity and number of live microbes inside a bottle of probiotic supplements does not match the label, never mind the sizeable issue of whether a given product can effectively help the consumer. So the bottle of *Lactobacillus ProHealthy* may in fact contain other types of bacteria not listed on the label, or may not even contain *Lactobacillus ProHealthy* at all. And my company, Immune Boosters, never has to provide a single piece of documentation proving that *Lactobacillus ProHealthy* is pro-health!

Because companies can benefit from selling a probiotic without demonstrating its effectiveness, there is little incentive to explore new potential probiotics. Therefore, probiotic availability is primarily limited to just a few groups of traditional types—those that have been consumed in fermented foods for ages. Although there are likely many other types of bacteria from a variety of environments that are good candidates for probiotics (including the human digestive tract), the lack of a history of safe consumption is a huge barrier to these becoming actual products. If I decide Immune Boosters should market a new probiotic supplement containing a recently identified bacterium that shows promising health benefits in research studies, even if I didn't want to include a specific health claim on the label, I would still need to prove that this new bacterium was safe to consume. That means undergoing extensive and expensive studies in animals and humans, or risking vulnerability to legal action by consumers or the FDA. Most companies have concluded that this gamble is not worth taking.

THE CLAIM GAME

The American probiotics industry has historically walked the fine line of making health claims that are just shy of those that would require a lengthy and expensive set of FDA-mandated clinical trials.

Dr. Sanders points out that in the United States, probiotic companies can make what is known as a structure-function claim, which relates the product to the normal "structure function" of the human body, without the approval of the FDA. Although these claims have to be truthful and not be misleading, the requirement for evidence is fairly relaxed. The Federal Trade Commission (FTC) is the organization in the United States that determines whether a product has adequate substantiation for any advertising claim it makes. In 2010 Dannon famously crossed this line when it claimed that consuming Activia was "clinically proven to help regulate your digestive system in two weeks." The FTC ruled that Dannon had gone too far in its health claim and sued the company for false advertising. Dannon has since dropped the term "clinically" and ads for Activia no longer mention that it can ease irregularity.

Companies selling probiotic products employ clever marketing tactics to convince you that their products will improve your health. But is it fair to say that all these claims are hype? Unfortunately, the science behind the effects of probiotics on the intestinal microbiota and host health has been partly undermined by pseudoscience resulting from poorly performed studies, many of them funded by probiotic and yogurt companies with obvious conflicts of interest. However, as our understanding of the microbiota improves, the role that probiotic bacteria play in our health is becoming an area of more serious scientific inquiry. Mary Ellen Sanders remains hopeful about the future clinical use of probiotics. "There is compelling evidence for some clinical benefits of probiotics, which some clinical organizations have embraced." As more rigorous studies are undertaken to determine how probiotic bacteria can benefit our health, probiotics will become a more accepted avenue to impact our biology.

PREBIOTICS AND SYNBIOTICS: PROBIOTIC SHELF-MATES

Prebiotics, unlike probiotics, are not living organisms, but as with probiotics, the ultimate goal in ingesting them is to increase the number of good bacteria in your colon. Prebiotics are food-derived compounds, usually long chains of linked sugar molecules known as complex carbohydrates or polysaccharides—a purified form of dietary fiber—that are not absorbed or metabolized by the host (us) and therefore provide nourishment to the bacteria in the colon. Upon arriving in the colon, prebiotics can be fermented by bacteria within the microbiota to promote their growth and abundance, and positively impact health.

One of the most common commercially available prebiotics is inulin, a polymer of up to sixty molecules of fructose that are joined together like links on a chain. Inulin can be purchased as a dietary supplement, but it is naturally found in many fruits and vegetables (especially bulbs such as onions and tubers like Jerusalem artichokes). With the justified backlash against consuming large amounts of fructose—usually in the form of high fructose corn syrup—it may seem counterintuitive to consume large polymers of fructose with the goal of improving health. However, in this case the devil is in the details. The fact that inulin is a polymer of fructose changes its fate in our digestive tract from that of a single molecule of fructose found in corn syrup. Like a sponge in water, our digestive system is very adept at absorbing single fructose molecules and shuttling them into our bloodstream. Bacteria are also good at fermenting fructose, but because we absorb it so early in the digestive process, little, if any, of it ever reaches the microbes in our large intestine. In contrast, the human genome does not encode the capacity to snip the chemical bonds that link fructose together in inulin, so these linkages function like a locked cage, making fructose molecules unavailable to us. Once this "cage" reaches the microbiota, which is filled with bacteria that have the key to unlock it, the doors of the cage are opened and the microbiota can feast on its contents—the single fructose molecules. If we didn't have a microbiota, inulin would pass straight through us and emerge virtually unaltered.

Gut bacteria ferment inulin and produce short-chain fatty acids. As we mentioned in Chapter 3, SCFAs can be absorbed for energy and can protect our gut from inflammation. So while fructose has received a bad reputation, it's important to think about the form in which you are consuming fructose. In its polymeric form, like inulin, it can provide sustenance to the microbiota.

Many prebiotics are simply purified forms of dietary fiber and therefore are also found naturally in abundance in plants. For instance, inulin and fructooligosaccharides, or FOS, along with many other carbohydrate polymers, are abundant in onions, garlic, and Jerusalem artichokes. In fact, almost all plant-based carbohydrate polymers and dietary fiber can be thought of as prebiotics that can be consumed by members of the microbiota.

The entire produce section of the grocery store should have signs and stickers: "Contains Prebiotics!" Over the course of studying the microbiota our family has adjusted what we eat to maximize produce and legumes, largely for their prebiotic content. As with probiotics, solid scientific evidence for the health benefits of individual prebiotics is still emerging. However, much evidence supports the health benefits of increasing dietary fiber intake to support your microbiota.

Synbiotics are a combination of probiotics and prebiotics. The "syn" in synbiotics stands for synergism, since these combinations are supposed to exert an effect that is greater than the sum of their parts. Prebiotics provide food for the probiotic, allowing the bacteria that you consume to become more abundant once they reach the colon. Like probiotics, synbiotics are not drugs and are not regulated by the FDA and cannot claim to treat disease. So they often have suggestive names, like Floramune, and their labels are worded very carefully, for example, "helps restore the balance of the intestinal microflora." Synbiotics are also becoming more widespread in stores, but we commonly make our own synbiotics by having a bowl of yogurt (probiotic) with banana slices (inulin-containing prebiotic) on top. Or we top a salad containing onions (prebiotic) with a dressing made from sour cream or kefir (probiotic). Keep in mind that most fruits and veggies are great sources of prebiotics.

THE FUTURE OF PROBIOTICS

With the refinement of specific strains, in the future probiotics may help to treat irritable bowel syndrome, inflammatory bowel disease, and even obesity and obesity-related conditions. "I think there will be probiotic drugs out there, and they will cost ten times as much as your yogurt will. They'll have stronger claims associated with them, but they might be the same exact stuff," says Sanders. But until then, concrete recommendations about which types of probiotics work best for which conditions are still a ways off. Considering the individuality of each person's microbiota, studies will likely benefit from extensive characterization of each subject's microbiota before, during, and after a probiotic study. If only 10 people in a 100-person clinical trial show a given response to a probiotic, the probiotic is likely to be considered ineffective. However, if those 10 responders all share a similar microbiota fingerprint that is distinct from the 90 nonresponders, then perhaps we could predict who would be most likely to benefit from that probiotic. But while we're waiting for the promise of personalized medicine to hit the neighborhood clinic, it's likely that the future of probiotic treatments will include bacteria other than those isolated from fermented dairy products. Perhaps they will come from the stool of humans. In fact, many *Bifidobacterium* strains found in probiotic supplements or in yogurts were originally isolated from the diapers of healthy infants and used historically to treat diarrhea.

The explosion of information about the inhabitants of the human microbiota is offering clues about what types of bacteria might be effective new probiotics. *Faecalibacterium prausnitzii*, a species of bacteria that is commonly found in the human gut, is often depleted in individuals suffering from inflammatory bowel disease, Crohn's disease, ulcerative colitis, and colorectal cancer. Mice harboring this bacterium have decreased gut inflammation and other positive immune system markers, making it a good candidate for an effective probiotic. Only time will tell whether certain bacteria, like *F. prausnitzii*, that are depleted in diseased states can be reintroduced to help improve symptoms. Clearly there is a huge health potential in supplementing beneficial bacteria that aren't just transient members, but could take up permanent residence in the gut.

We may also come to rely on the use of multiple types of bacteria, a probiotic cocktail of sorts. Bacteria can participate in synergistic relationships with one another or with the humans they inhabit, so combining compatible strains can lead to an effect that is greater than anticipated. If the gut contains an unstable microbiota, throwing in just one bacterial strain may be the equivalent of having unequipped firefighters show up to a six-alarm blaze. But if you combine complementary tools like ladders and fire hoses, and include other rescue workers, their collective action can be highly effective. Raking though the types of bacteria found in healthy microbiotas may help to identify particularly beneficial strains. Combinations that play nicely with one another are a potential path to new probiotics.

But outside our gut there is another dense collection of microbes that may contain a probiotic diamond in the rough. Dirt.

Geophagia, the consumption of dirt, is widespread in the animal kingdom. We've all consumed dirt inadvertently from our hands or from insufficiently washed produce, and consumed it purposefully in the form of table salt. In some human cultures around the world, dirt is consumed consciously. In Haiti some people eat *bon bons de terre*, which are earth cookies made with butter, sugar, and dirt. According to the *Diagnostic and Statistical Manual of Mental Disorders*, dirt consumption is abnormal, even though it has been practiced for hundreds of years.

Other than in cases of starvation, it's not clear why some humans have a craving for dirt. Most theories revolve around the idea that it can satisfy a nutrient deficiency or that the dirt, clay specifically, can absorb toxins out of the digestive tract. In fact, geophagia can be an effective treatment for nausea. But what if, in addition to adding nutrients and minimizing toxin exposure, consuming the microbes in dirt is beneficial for humans? One company is making this argument and marketing a probiotic supplement in which the bacteria, instead of being from fermented dairy products or human isolates, is a mixture of microbes normally found in dirt. There is some evidence that the consumption of soil bacteria can ameliorate symptoms associated with irritable bowel syndrome. Perhaps the lack of dirt in the diet of people living in the hygienic industrialized world is problematic, and probiotics from soil offer a way to recapture evolutionarily important interactions. Whether or not consuming soil bacteria will stand up to the rigors of scientific scrutiny is not yet known, but probiotics made from soil bacteria may be worth a try if probiotics from more traditional sources don't seem to provide a benefit.

One exciting possibility for future probiotics is the genetic engineering of bacteria. Imagine a scenario in which you are experiencing inflammation in your gut as a result of inflammatory bowel disease. What if there were an engineered "smart" probiotic which, as it transited your gut, could sense exactly where there was inflammation and deliver a targeted anti-inflammatory molecule directly to that site—a smart bomb of the microbe world. The probiotic could then sense when the inflammation was under control and stop releasing the anti-inflammatory drug. Bacteria could also be engineered to perform diagnostic tests, serving as sensors that could detect disease in its earliest stages. These microbial beacons might even make the dreaded colonoscopy obsolete.

A PROBIOTICS USER'S GUIDE

Our ancient ancestors were bacteria eaters on a scale that most of us do not approach. Some of these bacteria were beneficial and others more problematic. It's the problematic bacteria that have led to the hypersanitization of our food and water supply, our homes, even our antibacterial-laden clothes, kitchenware, and plastic tchotchkes. While few would argue that eliminating as many pathogenic microbes as possible from our environment is a bad thing, it may be that our approach of complete microbe extermination is not the best thing. Perhaps instead of trying to create a microbial void we should be replacing bad microbes with beneficial microbes like probiotics.

Before using probiotics to treat a medical condition it is important to confer with your physician to determine which particular probiotics might be best for you. While probiotics have been consumed safely for hundreds of years, there can be concerns for immunocompromised patients, underlining the need to talk with your physician. Probiotics may be most beneficial in helping healthy people prevent disease rather than treating a medical condition.

Consumable, living bacteria are available in a variety of forms such as supplements, fermented foods that have not been sterilized (like yogurt, unpasteurized sauerkraut or pickles, kimchee, or miso paste), and nonfermented foods that have had live bacteria added to them (such as Goodbelly, a bacteria-supplemented fruit juice). Sour cream, butter, and some cheeses are examples of foods that may or may not be fermented and contain live bacteria. The bacteria-containing versions of these products commonly contain fewer microbes than required for the "live and active cultures" label (at least 100 million bacteria per gram), but often can be identified by the word "cultured" on the label. Other foods that have traditionally been fermented, like pickles, are now often pickled using vinegar brine, bypassing the use of bacteria. Some fermented foods are pasteurized, a process that kills the bacteria, and are therefore not a source of living microbes. If a fermented food is not refrigerated but stored at room temperature in jars or cans (e.g., canned sauerkraut) it likely contains no living microbes. So it's important to read labels carefully if you want to ensure that the product contains living microbes. Most probiotic products proudly display that fact on their label.

Our family consumes microbes regularly, usually in the form of fermented dairy products like yogurt and kefir. When an illness seems imminent, our bacteria consumption increases. Our preference for yogurt and kefir is purely a personal one and not a result of any knowledge that these probiotic bacteria are any better than those found in other types of fermented foods. We occasionally eat miso, kimchee, and even ferment our own pickles. But when choosing items from the yogurt section of your grocery store, be mindful of the sugar bombs parading as healthy kids' snacks. Unsweetened fermented dairy products can definitely be an acquired taste for some, especially children. The lactic acid bacteria used to ferment yogurt generate that sour, tangy taste that flavored yogurts curb with sweeteners. One way to keep kids from immediately rejecting the sourness of plain, unsweetened yogurt is to add your own sweetener, such as honey or maple syrup, and then gradually decrease the amount until it is no longer necessary. Adding fresh or frozen berries or other fruit is another way to sweeten plain yogurt and at the same time add some prebiotics.

Our family does not routinely consume probiotic supplements. We feel that the variety of bacteria found in fermented foods offers the best chance that we'll encounter a microbe that benefits our health. However, in the past we have used supplements, along with fermented foods, following a course of antibiotics, adding an even larger population of bacteria to the gut to compensate for the microbiota damage. The other scenario in which we would consider a supplement is after diarrheal illness. Antibiotic use and diarrhea are two circumstances that opportunistic pathogens can exploit to cause problems. The extra boost of bacteria that probiotics supply could potentially ward off a nasty microbe trying to take advantage of the vulnerable state of your gut.

Because of the individual nature of each person's microbiota and the inability to predict which type and how much probiotic might be helpful and for what conditions, it is important to find probiotics that work well with your microbiota. Any probiotic product that causes uncomfortable bloating, excessive gas, or headaches is not a good match for you. One of the more obvious benefits you should experience with probiotics is more regular and easily passed stools. It may require a little trial and error with various types of probiotic-containing foods or supplements to find one that agrees best with your system.

There are a number of probiotic foods you can experiment with, many of which are dairy products, but there are also nondairy options. We have provided a list of fermented foods in the Appendix to help you navigate the possibilities. There are also great online suppliers of starter cultures to help you produce your own yogurt, kefir, kombucha, and even fermented soy products, rice, and vegetables. One example of a company that we have used is Cultures for Health. If none of these options are right for you or you feel that you would be better served with a supplement, remember that you have many options derived from different sources. To avoid potentially shady probiotic manufacturers it is important to buy from trustworthy companies. Most reputable probiotic companies provide information about studies conducted with their products and have labels that clearly provide the names of the bacteria contained within, as well as information on their shelf life. Products that only specify the date of manufacture should be treated with suspicion. Organizations like the U.S. Pharmacopeial Convention (USP), a scientific nonprofit organization, provide third-party product evaluation indicated on the product's label.

In the search for the right probiotic, it is important to systematically try different ones until you find something that seems to work well for you. How can you tell? In the absence of symptoms that you are trying to eliminate, the biggest clue we have about what is happening to the microbiota is your stool. The ideal stool is smooth, soft, and easy to pass and comes out in one long snake-like piece without any cracking, which is an indication of constipation. The lack of a splash means you are on the right track.

CHAPTER 5

Trillions of Mouths to Feed

THE MICROBIOTA EXTINCTION EVENT

As the human diet has transitioned from hunting and gathering to farming and now to the consumption of factory-produced foods, the microbial community in the gut has had to adjust. Over this period of technological innovation in food production, some species of bacteria have gone missing and are potentially extinct in the guts of modern Westerners. There are many factors driving microbiota diversity loss. Part of this microbial extinction is due to the lack of food-borne microbes (the good kind), which can be restored by eating fermented foods, as discussed in the previous chapter. A second factor is the lack of fibrous plant material in our diet. Plants have nurtured a diverse microbiota in humans for millennia and now that they are no longer a large part of our diet, our microbes are suffering.

These two methods of improving microbiota diversity—the increased consumption of beneficial microbes and an improvement in the food we provide to our gut-resident microbes—can be used simultaneously to reverse the Western microbiota extinction event occurring in our gut. Sources of "extra" bacteria include fermented foods with bacteria, such as yogurt, pickles, sauerkraut, kimchee, and kombucha, as well as environmental microbes from gardens and pets. Not sterilizing our homes with toxic antimicrobial cleaners is also likely to help introduce microbes to the gut. But as you increase your exposure to microbes in food and your environment, diet is an important factor in determining which microbes will stick around over time.

Increasing dietary fiber is essential to cultivating diversity in the microbiota. Microbes in the gut thrive on the complex carbohydrates that dietary fiber is primarily composed of. These complex carbohydrates are vastly different from the appropriately vilified simple carbohydrates found in starchy food and sodas that are absorbed in our small intestine and rarely reach the microbes living farther down, in the colon. But rather than "dietary fiber," an imprecise term, we prefer "microbiota accessible carbohydrates," or MACs. MACs are the components within dietary fiber that gut microbes feed on. Eating more MACs can provide more nourishment to the microbiota, help gut microbes thrive, and improve the diversity of this community. But this requires a massive change in the fiber-sparse eating habits of the industrialized world. Our family eats what we jokingly refer to as a "Big MAC diet." This diet is rich in complex carbohydrates from fruit, vegetables, legumes, and unrefined whole grains, and is designed to create and maintain diversity within the gut microbiota.

OUR MICROBIOTA: THE ULTIMATE RECYCLERS

Embedded in the numerous revelations about how the gut microbiota controls aspects of our health, there is a central theme that reveals how we can exert control over these microbes: the microbiota is directly responsive to diet. Both the membership of the gut community (also referred to as its composition) and what they are doing—their functionality—are a direct result of what you eat. What choices can you make to create and support the best microbiota possible? In a life full of difficult decisions about what to eat—Is a low-fat or low-carb diet better? Should I be eating organic food? Can I make up for the massive mound of French fries I just ate by leaving a few on my plate?—there are some simple rules that you can follow to increase the MAC content of your diet and foster a healthy microbiota. But to make the microbiota healthy, we need to understand how to feed these microbes, and that requires some basic knowledge about what happens as food passes through our digestive tract.

The digestive system operates much like an efficiently operated waste management facility. Just as discarded items are dumped onto a conveyor belt for sorting, the stomach unloads its contents (our latest meal) into the small intestine. The digestive tract begins a process of sorting through a diverse array of material—fats, proteins, carbohydrates, salt, vitamins, and numerous other compounds. On the conveyor belt, valuable materials like glass, metal, and other recyclables are picked off first; similarly, the small intestine absorbs valuable "recyclable" material such as simple carbohydrates, amino acids from proteins, and fatty acids. These food components have high caloric value and can be easily used for energy or, in some cases, are actually recycled by our own cells to build new tissue.

Next in waste management is the removal of biological material that can be composted. Similarly, the leftover indigestible, unabsorbed portion of our last meal moves into the large intestine for microbial transformation. Much of the material destined for the colon is dietary fiber, which our human enzymes within the small intestine are incapable of digesting into useful calories or nutrients.

However, for the microbiota, this dietary fiber is loaded with MACs and provides a veritable microbial feast.

THE VALUE OF MICROBIAL WASTE

The microbes in our gut are completely dependent on our food choices. Some species prefer to eat the MACs within bananas, while others have better success with those from onions. Depending upon what we eat, we dictate which of these microbes will be successful, multiply quicker, and thereby become more abundant. On the other hand, the components of food that these microbes specialize in eating, complex carbohydrates, are not calories that are accessible to us. These microbes are not freeloading, rather just depleting material that would go unused by our body.

Like all life forms on Earth, microbes need to absorb and metabolize molecules to provide energy for their own growth and reproduction (technically called cell division for bacteria). For species like ours that sexually reproduce, it may seem somewhat egotistical, but the goal of each bacterium is to make as many copies, or clones, of itself as possible. Those species that are most effective at reproducing in a given environment will persist and dominate—an example of natural selection in its most basic form. When combined with a bacterial genome's ability to acquire, delete, or modify genes, over generations microbes can evolve and hone their ability to compete effectively in the gut.

The competition for nutritional resources in the gut is intense, forcing microbes to evolve diverse and clever metabolic strategies for survival. But regardless of the strategies employed, all gut microbes face a couple of big challenges to obtain calories. The first is how to extract energy in the absence of oxygen. The gut is an oxygen-free, or anaerobic, environment. Human cells use oxygen to perform aerobic metabolism to generate molecular building blocks for cells and fuel for our bodies. However, the microbiota must rely upon anaerobic metabolism, or fermentation, to generate energy and create important molecules without oxygen. The second challenge these microbes face is the speed at which this metabolism must take place. Food moves through our digestive tract rapidly, and a competitive ecosystem forces bacteria to quickly consume any available passing nutrients. The strategy utilized by the most prevalent human gut-residing bacteria to solve these problems is rapid fermentation of MACs, one of the most abundant forms of energy within the colon. Microbes in a variety of environments outside the gut similarly perform fermentation. For instance, the bacteria used to make yogurt ferment the lactose in milk to lactic acid. In the most well-known type of fermentation, yeast metabolizes starch, sucrose, and other sugars into ethanol to make beer and wine. Unlike beer and yogurt, ethanol and lactic acid are rare end products of the fermentation that takes place in the gut. The most commonly manufactured fermentation products in the gut are short-chain fatty acids (SCFAs).

SCFAs provide humans with a small amount of calories salvaged from plant carbohydrates which, without microbial digestion, would have no caloric value. The microbiota squeezes every last bit of calories from the carbohydrates they feast upon, but can't generate any calories from SCFAs. Oxygen is required to generate calories from SCFAs, and the gut is an oxygen-free environment. As we absorb SCFAs from our gut into our own oxygen-containing tissue, our bodies wring out the last remaining calories from the otherwise indigestible fiber. On the ancient savanna, where calories were scarce, humans consumed much dietary fiber in the form of wild berries and roots. The resulting SCFAs the microbiota provided were likely an important contributor to their daily caloric requirements and may have been critical in furnishing enough energy for them to hunt and gather.

However, in the current calorie-filled Western diet, microbiota-generated SCFAs contribute only 6 percent to 10 percent of our total daily calories, the equivalent energy provided by about twenty almonds. On the whole it is not that much, but these are still extra calories for a population in which obesity-related diseases are reaching disastrous levels. Shouldn't we be trying to eliminate any and all unnecessary calories? What if we wiped out the microbiota? Would a sterile gut make us thinner? Maybe. Microbiota-free mice eat more food but weigh less than mice with a microbiota. But since humans can't permanently inhabit a sterile bubble (which is how these microbiota-free mice live), trying to eradicate the microbiota would require constant high levels of antibiotics. And even that would likely not be enough to keep our gut microbe free. Bacteria are highly adaptable; antibiotic-resistant bacteria would rapidly fill our gut.

While SCFAs provide a few extra calories, it is becoming clear that they actually serve a much more important role throughout our body. There is an emerging view that we should, in fact, be attempting to boost SCFA production by consuming more MACs. SCFAs are important mediators of functions within our body, and there is evidence that they don't contribute to weight gain.

NOT JUST EMPTY CALORIES

The list of ways in which the SCFAs influence our health is growing. SCFAs provide extra calories, yet people consuming a SCFA-generating high-fiber diet actually lose weight. This paradox is reminiscent of the one observed in France, where people consume a diet relatively high in fat but don't gain weight as readily. One possible explanation is that SCFAs make us feel full longer, so we consume fewer calories overall. Maybe the production of SCFAs from fermentation of the spinach salad we just ate adds a few extra calories, but they may also make us feel sufficiently satisfied to resist reaching for a cookie or other dessert.

SCFAs are just one of potentially several types of health-promoting compounds that the microbiota can produce. The metabolic pathways encoded within the microbiota are extremely complex and are capable of synthesizing a variety of chemical molecules inside the gut. Scientists have detected a diversity of microbiota-produced molecules, but for many of them their identity and how they impact the body is unknown.

The idea that the lack of dietary fiber has changed the modern microbiota and explains much of Western disease is currently one of the prevailing theories in microbiota research. Providing more dietary fiber for microbiota fermentation would likely result in weight loss, lower inflammation, and decreased risk of Western diseases, not to mention a more stable, diverse microbiota. Many traditional societies consume much more plant material than modern Westerners do, and the plant material consumed in the West is often low in fiber, containing primarily simple starches that get "recycled" in the small intestine. Notably, individuals from high-fiber-consuming societies have a larger variety of bacteria in their microbiota (some of which have never been observed in the microbiota of Westerners) and much lower rates of inflammatory diseases. But is the concept of higher dietary fiber consumption resulting in less disease really new?

THE LONG-FORGOTTEN BENEFITS OF FIBER

Determining how the microbiota can be manipulated for better health is at the forefront of research in this field and diet is one of the most powerful levers we have to control the microbiota. However, almost one hundred years ago the *Journal of Medical Research* published a research study entitled "The Regulation of the Intestinal Flora of Dogs Through Diet." Even a century ago there was an understanding that the chemical characteristics of what the dogs ate—for example, the types of carbohydrates—mattered for shaping the membership of the microbiota. Perhaps such an early understanding of the impact of diet on the canine microbiota had to do with the fact that humans are often ultimately responsible for cleaning up Fido's excrement. With such an early understanding of the connection between diet and the microbiota, why have we been so slow to realize that the types of carbohydrates we consume impact our microbiota just as it does our dog's?

Dr. Thomas Cleave was one of the first modern physicians to promote fiber consumption in the 1950s. In his book *The Saccharine Disease* (1974) he formulates the argument that many modern illnesses are a result of the overconsumption of refined carbohydrates and the decreasing consumption of dietary fiber. Cleave was a British naval physician who cared for sailors during World War II. Because of the scarcity of fruits and vegetables on navy warships, these sailors often battled constipation, which Cleave remedied by prescribing bran. Their immediate recovery emboldened Cleave to advise the consumption of bran for many types of health issues ranging from diverticulitis and hemorrhoids to even cavities and headaches. His belief in the healing powers of bran earned him the nickname "Bran Man" as well as a reputation for being an overzealous proponent of fiber consumption. Many found the idea that modern diseases were a result of too much sugar and too little dietary fiber ridiculous, and Cleave was chided rather than embraced by the medical community.

Dr. Denis Burkitt was a surgeon who had spent a lot of time in African hospitals investigating and treating a type of cancer that became known as Burkitt's lymphoma. Burkitt had read some of Cleave's work and because of his experience in Africa had noted that the high-fiber diets of many Africans seemed to protect them against diseases like diabetes, heart disease, colorectal cancer, and even maladies like hemorrhoids and constipation. Like Cleave, Burkitt became extremely interested in the role that dietary fiber played in human health. Work by Burkitt and several others, including Alec Walker and Hugh Trowell, found that rural Africans passed stool that was three to five times more massive, had a gut transit time that was more than twice as fast, and ate three to seven times more dietary fiber (60 to 140 grams versus 20 grams) compared to their Western counterparts. Burkitt spent the rest of his academic life studying and extolling the importance of fiber consumption for health. His belief in the importance of a high-fiber diet is evident with his statement that, as a country, "if you pass small stools, you have to have large hospitals."

The work by Cleave, Burkitt, Walker, Trowell, and many others led the FDA in 1977 to recommend that Americans increase their consumption of dietary fiber. Food manufacturers followed suit by prominently displaying the dietary fiber content on their products. In 1997 the FDA allowed foods containing certain types of fiber to proclaim that they "may reduce the risk of heart disease." Cities where Burkitt publicly spoke about the link between fiber and health would quickly sell out of bran, and roughage became a household word.

So why aren't we all eating a high-fiber diet now? Unfortunately, around the time that the virtues of fiber were being sung, many people shifted their focus to dietary fat. Fat was deemed the enemy not only to our waistlines but also to the functioning of our heart and our propensity to a litany of modern diseases. Low-fat products popped up everywhere, and instead of looking at the grams of dietary fiber on packaging, Americans zeroed in on the number of fat grams. The low fat argument made sense on an intuitive level. If you wanted to lose fat, you ate less fat. It seemed so simple. The argument for high fiber was more cloudy—high fiber reduces the risk for Western diseases, but we didn't really know why.

In the foreword to Cleave's book *The Saccharine Disease*, Burkitt admits that while the link between low dietary fiber and Western diseases is clear, the reason behind this connection is not. "The mechanisms . . . to explain how dietary changes might cause various diseases may require modification in the light of advancing knowledge." The difference now is that we are finally beginning to understand why. Our microbiota needs fiber.

CARBOHYDRATES' BAD REPUTATION

The word "carbohydrate" comes with a lot of baggage. Each of us probably knows someone who has been on a low-carbohydrate diet, or we have been on one ourselves. Atkins, South Beach, Zone, Paleo, and other diets have proliferated rapidly and were even blamed by Krispy Kreme doughnuts for profit losses. But before we villainize carbohydrates, it is important to have a clear understanding of what they are. Carbohydrates are a group of organic compounds containing carbon, hydrogen, and oxygen that serve as the major energy source for animals. The collection of compounds that fall under the umbrella of carbohydrates is extensive. For our purposes, carbohydrates can be divided into three broad categories: those that are digested by the human, those that are digested by the microbiota, and those that pass through undigested.

First let's consider the carbohydrates that are digested by humans and absorbed in the small intestine without the help of microbes. Monosaccharides are the simplest carbohydrates, containing a single molecule of sugar such as glucose or fructose. Monosaccharides can be absorbed directly from the digestive tract into the bloodstream. Two monosaccharides linked together are called disaccharides; lactose and sucrose (table sugar) are examples of disaccharides. The monosaccharide and disaccharide content of food is listed on the nutritional label as "sugars." Polysaccharides are many monosaccharides linked together and qualify as complex carbohydrates. Starches are one type of polysaccharide. However, like the simple mono- and disaccharides, most types of starch are digested and absorbed before reaching the large intestine. Starch dominates many staples of the modern diet: pasta, white bread, potatoes, and white rice are all packed with starch. Most of this starch gets converted to the simple sugar glucose, is absorbed into the bloodstream before it reaches the microbiota, and is metabolically similar to the same quantity of sugar. Yet, the starch content in foods is not apparent when reading a nutritional label.

The second category of carbohydrates—the MACs—feed your microbiota. There are thousands of different types of microbiota accessible carbohydrates found in the different plants humans eat. Oligosaccharides consist of three to nine monosaccharides and are found in beans, whole grains, and many fruits and vegetables. Most oligosaccharides are not digested in the small intestine and pass to the colon, where they are rapidly fermented by the bacteria there. Similarly, nonstarch polysaccharides, like the pectin found in fruits and the inulin found in onions, are composed of ten to hundreds of monosaccharides linked together and are destined for conversion into short-chain fatty acids by the microbiota.

The final category is composed of the carbohydrates that pass through the digestive tract unaltered. Most of these are polysaccharides and possess some chemical or physical trait that makes them resistant to human or microbial digestion. Cellulose, the woody fiber found in plant cell walls, is one such recalcitrant polysaccharide. Although gut microbes of other organisms, such as those that reside in the cow rumen or termite gut, are accomplished cellulose digesters, this task requires much more time than is available in a typical transit of the human gut.

It is the simple sugars used for sweetening and the easily digested starches that have given carbohydrates a bad name. These "bad" simple carbohydrates cause blood sugar to spike soon after consumption. The body responds to high blood sugar by releasing insulin, allowing cells in the liver, muscles, and fat to absorb the circulating sugar. Insulin also prevents the body from using fat as energy until all the sugar has been either used up or stored away in the form of glycogen.

If blood sugar levels are constantly elevated, say from a diet high in simple carbohydrates, insulin-responsive cells become resistant to insulin. These cells become desensitized to the constant high levels of insulin and begin to ignore it—a common step in the progression of type 2 diabetes. The outcome of this desensitization is dangerously high blood glucose levels that can result in heart disease, stroke, and kidney failure.

How quickly the carbohydrate within a particular food makes our blood sugar surge after eating it is measured by that food's glycemic index. The monosaccharide glucose, which has the fastest absorption into the bloodstream, has a glycemic index of 100. A food's glycemic index can be categorized into high (greater than 70), medium (between 56 and 69), and low (less than 55). The more easily digested carbohydrates (like the mono- and disaccharide types) a food contains, the higher its glycemic index will be. White bread, white rice, and potatoes are all examples of foods with a high glycemic index. Medium glycemic index foods include whole wheat bread, brown rice, and unpeeled potatoes. Beans, seeds, and intact grains have among the lowest glycemic index, owing to their minuscule amounts of monosaccharides, disaccharides, and starch and their abundant nonstarch complex carbohydrates.

More important than the glycemic index is a food's glycemic load. Glycemic index tells you how fast the carbohydrate in a food will raise blood sugar levels. Glycemic load takes into account the amount of carbohydrate within a certain quantity of food—for example, a serving—that will cause your blood sugar to rise. Pumpkin provides a good example of how glycemic load is more instructive than glycemic index. Because of the types of carbohydrates found in pumpkin, it technically has a high glycemic index. But the overall impact that a serving of pumpkin will have on your blood sugar level is very small, as reflected in its low glycemic load. Most vegetables have low glycemic loads and high MAC content. Steamed or boiled (or even microwaved, in a pinch) edamame, fresh fruit and nuts on yogurt, and whole grain bread with hummus are some of our favorite low-glycemic-load, high-MAC snacks. We have found that using online resources to become familiar with foods and snacks, especially, that have a low glycemic load per serving is a good way to guide what we purchase at the grocery store.

READING NUTRITIONAL LABELS FOR YOUR GUT MICROBES

Contemplating the nutritional value of a food item in the grocery store can be overwhelming. By studying all the words on the packaging—the health claims, ingredient list, and nutritional facts—we hope *something* will clearly tell us whether we should or shouldn't buy this food. It can be hard to discern whether the health claims on the label, while the easiest to understand, are really instructive or just marketing hype. And even with our degrees in biochemistry, the ingredient list of many products can be incomprehensible—usually a clue that it's best to put that item back on the shelf.

The Nutrition Facts Label, a requirement by the FDA, is designed to provide simple and uniform information about the food it describes. The main components of this label include calories, fat, cholesterol, sodium, protein, and total carbohydrate content. Most people zero in on the number of calories and grams of fat or sugar and bypass the rest. Unfortunately, the two pieces of information about the carbohydrates in a food that we view as most important—glycemic load and quantity of microbiota-nourishing carbohydrates—are not available on food labels. In the absence of this information, understanding how to interpret the carbohydrates category will help inform you about the approximate amount of microbiota food present in the package.

Total carbohydrate is determined by weighing a sample of the food and subtracting the weight of protein, fat, moisture content, and ash (a measure of nonorganic molecules like iron and bicarbonate). In other words, carbohydrates are not measured directly but determined by what is left over after other components have been measured. Within the total carbohydrates category, subcategories of sugars and dietary fiber are typically listed on the label. You may have noted that sugars plus dietary fiber does not necessarily equal the amount of total carbohydrates, since several types of carbohydrates are not accounted for in these two subcategories. Sugar content is the weight of all the mono- and disaccharides, the carbohydrates that are easily absorbed into the bloodstream. Dietary fiber is a mixture of polysaccharides and serves as an indicator of whether a food will nourish your microbiota, but in this regard consumers should be aware of some important limitations.

The term "dietary fiber" has been given multiple definitions by different official groups. Some of these definitions include fermentation by the microbiota, similar to our definition of MACs, while other definitions are indifferent to the microbiota. The potential confusion about the term "dietary fiber" is compounded by the lack of standard methods used to measure dietary fiber in food.

According to the Food and Agriculture Organization of the United Nations, at least fifteen different methods have been used to determine dietary fiber for food composition labels. The different laboratory tests that are currently employed can result in slightly different fiber amounts. Better methods will eventually be developed to determine which carbohydrates in a food are likely to result in microbial fermentation in the colon, i.e., qualify as MACs. But remember that due to the differences in microbiota between people, and changes in the microbiota that occur over time, such a test will still only be an estimate. Until a microbiota-specific test is developed that quantifies the MACs present in a food, dietary fiber content serves as the best available approximation.

Regardless of the issues with defining and measuring dietary fiber, if you examine the Nutrition Facts Labels of many packaged foods, you will find that dietary fiber is lacking in much of what we normally eat. Packaged foods made with refined flour and copious amounts of added sugars provide no sustenance for the microbiota and likely translate into guts populated by starving microbes. The FDA recommends that an adult male consume 38 grams of dietary fiber per day while a woman should consume 29 grams. Despite these recommendations, the average American consumes a measly 15 grams of dietary fiber per day, a deficiency that is undoubtedly contributing to the malformation of the Western microbiota.

While images of emaciated microbes may be floating through your mind, this is not strictly the case: bacteria can be extremely resourceful in their dietary-fiber-deprived state. That is because they have another source of carbohydrates, our intestinal mucus. During times of low fiber consumption, gut bacteria can sustain themselves on the carbohydrates that our intestinal cells continually secrete into the gut environment, which serves as a barrier to protect our own human cells from direct contact with the microbiota. But by feasting on mucus carbohydrates, our microbes deplete the protective gut mucus layer, compromising barrier function and increasing inflammation. While the long-term effects of less gut mucus on human health are still unknown, preliminary experiments suggest that loss of intestinal mucus can lead to colitis. But the microbiota is very adaptable: provide sustenance in the form of dietary fiber and many microbes will switch their focus from eating your mucus to eating your most recent meal.

MACS

Because the term "dietary fiber" comes with uncertainties, we prefer microbiota accessible carbohydrates, or MACs, to think about how the food you eat feeds your microbiota. As discussed earlier, MACs are the carbohydrates found in a variety of plants, like fruits, vegetables, legumes, and grains, that are fermented by the microbiota. Dietary fiber within food, as well as dietary fiber supplements, can contain carbohydrates that are not accessible to the microbiota and are therefore not fermented. These nonfermentable fibers can be very effective at easing constipation, they serve as bulking agents that allow stool to absorb more water and result in an easier bowel movement. But to feed your microbiota and generate short-chain fatty acids, you need to consume MACs. The more MACs you consume, the more fermentation you will have taking place in your gut and the more SCFAs you will produce. Which MACs you feed to your microbiota dictates the subset of microbes that will flourish, how many different types of bacteria make up your microbiota (how diverse the community is), and what sort of functions this community carries out. If you eat a lot of inulin-containing onions, microbes that are good inulin fermenters will become more abundant in your microbiota. Apples will encourage pectin-degrading bacteria, wheat bran will feed those microbes that can consume arabinoxylan, and mushrooms will help mannan-focused microbes bloom within your gut. We've named some prominent MACs associated with these foods, but every plant contains a diverse array of microbiota-feeding carbohydrates (as well as many that completely resist microbial degradation).

It is impossible to measure the amount of MACs in a food sample the same way something like protein content is measured. Due to the individuality of each person's microbiota, a MAC to one person may not be a MAC to another. In 2010 a group of scientists were studying a type of enzyme called porphyranase. This enzyme degrades a type of polysaccharide found in marine algae, culinarily known as nori, a common sushi wrapper and topping for many Japanese dishes. Not surprisingly, certain marine bacteria carry a gene for this nori-degrading enzyme in order to eat marine algae. What did come as a surprise, however, was the presence of these genes in the gut microbiota. Why would a gut microbe have enzymes that can degrade seaweed? The answer became obvious when the scientists found these genes in the collective microbiota genome, or microbiome, of Japanese but not in North American microbiomes. At some point in history the microbiota of seaweed-eating Japanese adapted to utilize this new food source. How did this happen? The most likely answer is that as people ate seaweed, they also consumed the marine bacteria living on that piece of seaweed—another example of how not sterilizing our food leads to exposure to beneficial environmental microbes. As these marine bacteria passed through the large intestine, they transferred genetic material to the gut resident bacterium. Suddenly a new functionality within the gut was born.

This example of microbiota seaweed consumption illustrates two important points about the microbiota. First, unlike the rest of the human genome, the microbiome is highly adaptable to its environment in a relatively short time frame. The choice of plants in the diet serves as one of the primary controls to change or maintain the membership of the microbiota. If the Japanese individuals with the nori-digesting microbiota stopped eating nori entirely, eventually that capacity would disappear. Second, despite the overwhelming number of genes in the microbiome, only genes that are used fairly frequently and serve a useful purpose to the microbes are maintained. The genes found in our microbiota come at an energetic cost to the microbes that have to maintain and replicate them each time they divide. To minimize extra effort, microbes keep their genomes relatively tidy, saving only useful genes.

A RICH VERSUS POOR MICROBIOTA

If the type and amounts of MACs in your diet impact the composition of your microbiota, it stands to reason that by reducing dietary

MACs, as Westerners have done, the microbiota has adjusted. In a study published in 2013, a multinational group of scientists looked at the number of genes in the microbiome of 292 Danish individuals. They found that these individuals could be divided into two groups, one with microbiomes containing many genes and the other with microbiomes containing relatively few genes—either a "rich" microbiome or a "poor" one. The rich group was more likely to have anti-inflammatory bacterial species in their gut and was more likely to be thinner. The poor group not only had more inflammation-associated species, like those seen in people with inflammatory bowel disease, they were also more likely to be obese, have a higher insulin resistance, and have a greater metabolic potential for generating procarcinogenic compounds. In other words, these poor-microbiome people had a profile that put them on the road for type 2 diabetes, cardiovascular and liver diseases, and cancer. Those with the rich microbiome had more genes involved in generating health-promoting SCFAs. And can you guess which group was more likely to gain weight over time? The poor ones. Clearly, having a rich microbiome is desirable, but how can this be achieved?

A similar study conducted in France also observed the poor versus rich microbiome divide among people. The French researchers queried their participants about diet and found that those with the poor microbiome consumed fewer fruits and vegetables (and thus fewer MACs) than the rich-microbiome group. But hope was not lost for the poor-microbiome group. When placed on a lower fat and calorie, but higher protein and fiber diet for six weeks, not only did people lose weight, they also increased the gene richness of their microbiome. As richness improved, so did other measures of health, including lower circulating cholesterol levels and less inflammation.

These two studies provide an important clue about why certain obese individuals are protected from developing diabetes, heart disease, and obesity-related diseases while others are not. Similarly, it may explain why some people develop these same illnesses without being obese, the so-called skinny fat people. The results of these studies show that the richness (or lack of richness) of your microbiome is a better predictor of Western disease risk than your weight. In the future your doctor, instead of measuring your body mass index (BMI), may be more interested in evaluating the state of your microbiome and, if it's deemed poor, may even prescribe a diet rich in MACs.

Another possible means to remedy a poor microbiome is to add more species (and their accompanying genes) to your microbiota. In 2013 Washington University's Dr. Jeffrey Gordon led a study to examine the microbiota from twins that were discordant for obesity, meaning one twin was lean and the other obese. When they transferred a relatively low-diversity, or poor, microbiota from an obese twin into mice, those mice got fat. A transplant from the higher-diversity, or rich, microbiota from the lean twin resulted in lean mice. Then they tested what would happen when they put the two groups of mice together in the same cage. Mice eat feces, so when housed in the same cage, the obese mice ate the feces (and all the associated bacteria) from the lean mice, and vice versa. This experiment addressed the question we asked earlier: does adding bacteria to a low-diversity microbiota increase richness and improve health? In Gordon's experiment the "lean" microbes took up residence in the obese-microbiota mice, increasing their microbiota richness and serving as an antiobesity factor.

But before you rush out to beg your thinnest friends for a microbiota transplant sample, here's the catch: the obese mice needed to be fed a diet high in fruits and vegetables and low in fat for the obesity protection to work. When the researchers set up the same lean- and obese-microbiota cohousing experiment, but fed the mice a diet high in fat and low in fruits and vegetables, the obese mice got fat and the lean-microbiota bacteria did not stick around. To increase the richness of bacteria that reside in the microbiota, just consuming more microbes won't work. We all eat microbes, and some inevitably come from other people's microbiota. But the health-promoting ones will not stick around unless we are eating a diet that helps them persist within the gut.

REFINING THE MACS OUT OF OUR DIET

Where did all the MACs in our diet go? The history of our consumption of wheat provides a perfect illustration. Today wheat has an image problem, but that wasn't always the case. Humans have consumed wheat for more than ten thousand years, in very large quantities in some cultures. Why has an ancient dietary staple been so vilified recently?

A kernel of wheat, or wheat berry, is made up of the endosperm, the bran, and the germ. The endosperm contains all the food, in the form of simple starches, to feed a newly growing wheat plant. The bran coats the outside of the wheat berry in a hard shell of fiber. The germ, a fat-filled reproductive organ that also contains fiber, germinates to create a new plant.

Thousands of years ago people began using millstones to grind wheat berries into a meal, bringing about the birth of flour. However, this stone-ground wheat would be unrecognizable next to the factory-produced flour available today. The Industrial Revolution brought about steam-powered mills, allowing for the production of flour on a significantly larger scale. But manufacturers struggled to keep flour fresh during the months it took to transport it from the mill to the consumer. To solve this problem, producers realized that if they removed the oily germ (the part that goes rancid) from wheat before milling, they could extend its shelf life almost indefinitely. What they didn't know was that by removing the germ, they were also removing a large amount of the dietary fiber, not to mention all the other healthful micronutrients that are found in wheat germ. Millers then realized that by removing the bran as well, they could provide consumers with white, fluffy flour—composed entirely of endosperm—that many people considered better looking, more palatable, and easier to bake with. Technology has provided us "rich man's flour" inexpensively, but our microbiota's diet has become poorer. As milling technology improved, wheat could be ground into ever finer particles until it came to resemble the ultrafine talc that fills the bags on grocery store shelves today.

It's clear how removing the bran and germ from wheat would reduce the amount of MACs, but why does the fineness of flour impact MAC availability? Doesn't a bag of whole grain wheat flour have the same amount of MACs as intact wheat berries? Not quite. Some MACs survive the journey to the microbiota because our human genome does not encode the capacity to degrade them. It's like a lock for which we do not possess the key. Some MACs, however, reach the microbiota because they are embedded in food particles that are too large for our bodies to digest before they reach the colon. It's a time issue. In this case we have the key to the lock, but the lock, in a sense, is hidden. These "hidden" carbohydrates contain linkages that the human genome encodes the capacity to digest, but because they are surrounded in a protective coating that needs to be digested first, they survive the journey through our gut relatively intact, providing fermentable MACs. If the flour is milled into a very fine powder our digestive enzymes have time to break down more of the carbohydrate linkages and absorb the resulting mono- and disaccharides directly into our bloodstream. If the flour is more coarsely ground, then our enzymes don't have adequate time to access all the carbohydrate linkages, leaving some of them intact for our microbiota. You may be eating the same amount of bread that your great-grandmother did, but because her bread was made from coarser milled flour that still contained the bran and germ, it contained more MACs than today's bread. A loaf of refined white flour bread, which is in many ways more like a cake your great-grandmother would have eaten, contains almost no MACs. Eating bread made from whole wheat flour will provide 2 grams of fiber per slice. But if you go even further and eat a cup of cooked unmilled wheat berries, you will get about 9 grams of fiber, a quarter to one third of your total fiber needs for the day.

In the San Francisco Bay area, sourdough bread is everywhere. Some of us even go so far as to make our own bread with a sourdough starter, which, like our gut microbiota, is a complex community of microbes that specializes, in this case, in eating flour. In order to maintain as many MACs as possible in our bread, we mill our own flour from wheat berries using a small hand-cranked grain mill. Using flour that has not been factory milled ensures that our flour maintains some coarseness, not to mention all its bran and germ. And while our bread is definitely heartier than the spongy, snow-white store-bought bread, the bran and germ give it an unmatched flavor. Making bread with a sourdough starter, rather than using commercial yeast, is a great way to decrease the bread's glycemic load because the microbes in the starter consume much of the simple carbohydrates in the flour. Plus, it provides the added fun of playing with a microbial community in the kitchen. Sourdough bread is also available commercially. Look for those that are made with whole grain flour for a higher MAC content.

WHAT ABOUT THE INUIT?

With all the evidence pointing to the benefits of a high-fiber diet, there is still quite a bit of skepticism from people who feel a high-protein diet is better. What about the Inuit? They eat almost no fiber and are very healthy. It's true that people living in the polar regions of North America traditionally ate almost no plants. Dietary fiber would be consumed only in the summer, when berries, tubers, or seaweed could be gathered. There is evidence that this dietary fiber reintroduction once a year was met with some discomfort. "The Angmagssalik Eskimos [will] get stomach pains from eating large quantities of seaweed after a long period without it. But after a few days of training they can again eat it without stomach pain," reported Kerstin Eidlitz in the 1969 book *Food and Emergency Food in the Circumpolar Area*. Perhaps the pain they experienced was the result of a sudden fiber-induced increase in fermentation, which in addition to producing SCFAs also produces gas.

Here, it is worth taking a moment to consider a question that many of you may be pondering. Won't a fiber-rich diet make me incredibly gassy? Eating healthier food can be difficult enough in our convenience-food-filled society. Why eat a diet that is also going to make us social outcasts! It is true that one of the by-products of bacterial fermentation is the production of gases like hydrogen and carbon dioxide. These gases are odorless, but as they exit the body, they carry with them a variety of malodorous volatile molecules produced by certain members of the microbiota, including some that contain sulfur (as in the smell of rotten eggs). Within the complex ecosystem of the gut, where one microbe's waste is another's food, the gaseous fermentation products of some species can actually be consumed by other microbes found in the gut, like *Methanobrevibacter smithii*. This microbe is not a bacterium but a member of the Archaea kingdom of single-celled organisms. *M. smithii* uses hydrogen and carbon dioxide gas to create methane, another odorless gas. Due to the biochemistry of making methane, *M. smithii* actually sucks up more molecules of gas than it produces. Therefore, *M. smithii* in your gut can help to reduce the total amount of gas to be expelled. With a diverse collection of microbes living in the gut, there is a better chance that gases produced by fermentation can be part of the complex food web and be consumed by other microbes. The more gas consumed by microbes, the less you have to expel.

The seasonal consumption of dietary fiber by the Inuit may have been adequate to allow their microbiota to regain richness annually. The few days of discomfort likely represent a period of microbiota adjustment to the new diet. The Inuit also consume quite a bit of fermented seal flipper, which while not supplying MACs would be a supplement of diverse microbes to their microbiota. In truth, however, because there have been no studies done on the microbiota of Inuits eating a traditional diet (which may now be impossible since Inuit acculturation is resulting in the Westernization of their diet), we may never know the makeup of their traditional microbiota. The other important fact to keep in mind is that because of their geographic isolation these individuals may have accumulated adaptations in both their own genome and that of their microbiome, allowing them to remain healthy on a meat- and fat-rich but fiber-depleted diet. Perhaps as with the effect of dietary nori in Japan, the Inuit microbiome accumulated genes allowing it to thrive in this extreme environment. These adaptations, assuming they had occurred, would not be present in the microbiota of the average Westerner. Much more information is required before the Inuit can serve as an effective counterargument for the necessity of increased MAC consumption.

Several studies show that a meat-centered diet impacts the microbiota in a way that is detrimental to health. Within four weeks, dieters on a high-protein, reduced carbohydrate regimen had a dramatic decrease in both the amount of SCFAs and fiber-derived antioxidants they produced and a buildup of hazardous metabolites in their colons. This type of environment would negatively affect long-term colon health by increasing the risk for inflammatory diseases and colon cancer. The microbiota of omnivores, compared to that of vegetarians and vegans, produces more of a chemical that is associated with heart disease. That compound, trimethylamine-*N*-oxide (TMAO), is a product of the microbiota metabolizing a chemical abundant in red meat. The growing evidence from these types of studies shows that a diet rich in MACs leads to a rich microbiota, which has health benefits for us.

THE "BIG MAC" DIET FOR A RICHER MICROBIOTA

Our family eats fish, dairy, and small amounts of pasture-raised meat, but the main portion of our plate is filled with MACs in the form of brown rice or cooked whole barley, beans, roasted vegetables, and usually fruit and/or dark chocolate for dessert. We limit our consumption of simple carbohydrates by staying away from factory-produced packaged foods and we limit the use of refined flours in baking. Getting enough dietary fiber can be difficult. But by cooking a lentil- or bean-centered dinner at least a couple of times a week and always having a jar of cooked beans on hand to sprinkle on salads or to make a quick quesadilla, we can easily boost our fiber intake. We prefer the taste of beans made from scratch as opposed to out of a can, so on the weekend we prepare a large pot of black beans, chickpeas, kidney beans, or whatever type we happen to have on hand. A slow simmer for a few hours only requires that we are around to keep the pot from boiling over or drying out. We store the cooked beans in glass jars in the refrigerator for the week or in the freezer for longer. We also frequently sprinkle nuts and seeds on whatever we are eating, including salads and main dishes.

If embracing such a diet sounds overwhelming, you may want to start with small steps to make it manageable. For example, get in the habit of starting each meal by assessing the dietary fiber content of the food on your plate. If you don't see anything that will provide food for your microbiota, think about key substitutions in how you prepared or ordered your food that could have improved the MAC content. If you worry about bloating and the discomfort that can accompany sudden changes in MAC consumption similar to the seasonal changes experienced by the Inuit, slowly increasing the amount of MACs in your diet over weeks or months will make the transition tolerable. Gradually you can begin to make choices at the grocery store and in restaurants to better nourish your gut microbes and enjoy better health as a result. We have provided recipes for MAC-filled dishes at the end of the book to help as you move up to a more microbiota-friendly way of eating.

CHAPTER 6

A Gut Feeling

THE BRAIN-GUT AXIS

A primal connection exists between our brain and our gut. We often talk about a "gut feeling" when we meet someone for the first time. We're told to "trust our gut instinct" when making a difficult decision or that it's "gut check time" when faced with a situation that tests our nerve and determination. This mind-gut connection is not just metaphorical. Our brain and gut are connected by an extensive network of neurons and a highway of chemicals and hormones that constantly provide feedback about how hungry we are, whether or not we're experiencing stress, or if we've ingested a disease-causing microbe. This information superhighway is called the brain-gut axis and it provides constant updates on the state of affairs at your two ends. That sinking feeling in the pit of your stomach after looking at your postholiday credit card bill is a vivid example of the brain-gut connection at work. You're stressed and your gut knows it—immediately.

The enteric nervous system is often referred to as our body's second brain. There are hundreds of million of neurons connecting the brain to the enteric nervous system, the part of the nervous system that is tasked with controlling the gastrointestinal system. This vast web of connections monitors the entire digestive tract from the esophagus to the anus. The enteric nervous system is so extensive that it can operate as an independent entity without input from our central nervous system, although they are in regular communication. While our "second" brain cannot compose a symphony or paint a masterpiece the way the brain in our skull can, it does perform an important role in managing the workings of our inner tube. The network of neurons in the gut is as plentiful and complex as the network of neurons in our spinal cord, which may seem overly complex just to keep track of digestion. Why does the gut need its own "brain"? Is it just to manage the process of digestion? Or could it be that one job of our second brain is to listen in on the trillions of microbes residing in the gut?

Operations of the enteric nervous system are overseen by the brain and central nervous system. The central nervous system is in communication with the gut via the sympathetic and parasympathetic branches of the autonomic nervous system, the involuntary arm of the nervous system that controls heart rate, breathing, and digestion. The autonomic nervous system is tasked with the job of regulating the speed at which food transits through the gut, the secretion of acid in our stomach, and the production of mucus on the intestinal lining. The hypothalamic-pituitary-adrenal axis, or HPA axis, is another mechanism by which the brain can communicate with the gut to help control digestion through the action of hormones.

This circuitry of neurons, hormones, and chemical neurotransmitters not only sends messages to the brain about the status of our gut, it allows for the brain to directly impact the gut environment. The rate at which food is being moved and how much mucus is lining the gut—both of which can be controlled by the central nervous system—have a direct impact on the environmental conditions the microbiota experiences.

Like any ecosystem inhabited by competing species, the environment within the gut dictates which inhabitants thrive. Just as creatures adapted to a moist rain forest would struggle in the desert, microbes relying on the mucus layer will struggle in a gut where mucus is exceedingly sparse and thin. Bulk up the mucus, and the mucus-adapted microbes can stage a comeback. The nervous system, through its ability to affect gut transit time and mucus secretion, can help dictate which microbes inhabit the gut. In this case, even if the decisions are not conscious, it's mind over microbes.

What about the microbial side? When the microbiota adjusts to a change in diet or to a stress-induced decrease in gut transit time, is the brain made aware of this modification? Does the brain-gut axis run in one direction only, with all signals going from brain to gut, or are some signals going the other way? Is that voice in your head that is asking for a snack coming from your mind or is it emanating from the insatiable masses in your bowels? Recent evidence indicates that not only is our brain "aware" of our gut microbes, but these bacteria can influence our perception of the world and alter our behavior. It is becoming clear that the influence of our microbiota reaches far beyond the gut to affect an aspect of our biology few would have predicted—our mind.

For example, the gut microbiota influences the body's level of the potent neurotransmitter serotonin, which regulates feelings of happiness. Some of the most prescribed drugs in the United States for treating anxiety and depression, like Prozac, Zoloft, and Paxil, work by modulating levels of serotonin. And serotonin is likely just one of numerous biochemical messengers dictating our mood and behavior that the microbiota impacts.

MICROBE-FREE MICE: GUTSY AND FORGETFUL

The idea that microbes can affect behavior is not that new. Many pathogens can influence the mind. The causative agent of syphilis, a highly motile, spiral-shaped bacterium called *Treponema pallidum*, can infiltrate the spinal cord and brain of infected individuals. Through its zombie-like possession of the nervous system, *Treponema pallidum* can induce depression, mood disorders, and even psychosis in its host. Some microbes even use mind control as a way to propagate themselves. The protozoa *Toxoplasma gondii* can infect rodents, travel to their brains, and cause them to lose their normal fear of cats, making them more likely to become prey. When a cat devours an infected rodent, *Toxoplasma gondii* benefits, using the cat to complete its life cycle, which ultimately disseminates through the cat's feces. In this case, *Toxoplasma gondii*'s "mind control" over the rodent is highly beneficial from the microbial point of view. The natural world is replete with examples of "bad" microbes like these using mind control over their hosts to their advantage. What is less known is whether "good" microbes can play the same mind tricks from the confines of the gut.

One of the first indications that the gut microbiota is wired into brain function and behavior came from scientists observing microbe-free mice housed in sterile bubbles. Scientists noted that microbe-free mice have a personality that is distinct from mice with a normal microbiota. Microbe-free mice are bigger risk takers and more inclined to explore their environment. In the rodent equivalent of extreme sports, these mice will travel longer distances in an open field, something that in the wild might get them more easily noticed by a hungry hawk. From an evolutionary perspective, a risk-taking mind-set may not be a good way to ensure survival or the ability to pass on genes to offspring. Shying away from an open field protects the mouse and increases the chance that its genes and its microbes will be passed on to future generations.

In their observations of these microbe-free gutsy mice, scientists discovered that if the mice were colonized with microbes, their behavior became more cautious, like that of the normal mice. But for this infusion of caution to work, microbes needed to be introduced before the mice reached adulthood. Once they were adults, adding microbes to their gut could no longer reverse the excessive risk-taking behavior. It seems that the role that gut microbes play in setting a mouse's tolerance to risk is confined to a critical period during infancy. In humans, infancy is a period of incredibly rapid brain growth and rewiring. If microbes play a role in the development of personality and behavior in humans, as they do in mice, it makes sense that gut microbes could exert their biggest impact during infancy.

Scientists have noticed that microbe-free mice are not just thrill seekers, they also have memory-related defects. A group of researchers put two groups of mice, one with a microbiota and one without, through some memory tests. In the first test the mice were given five minutes to explore two new objects, a small smooth napkin ring and a large checkered napkin ring. The objects were then removed for twenty minutes. After this rest period, the large checkered napkin ring was returned to the cage along with an object the mice had never seen, a star-shaped cookie cutter. If the mice remembered the napkin ring from before, they would pay less attention to it and instead spend their time exploring the unfamiliar cookie cutter. The mice with a normal microbiota did just that. The mice with no microbiota, however, spent just as much time exploring the "old" napkin ring as the "new" cookie cutter. These mice had completely forgotten an object they had just seen twenty minutes earlier.

It is important to remember (and maybe your microbiota will help) that the microbiota-free conditions used to perform these experiments in mice cannot be reproduced in humans—we're all colonized. However, by using these extreme conditions—no microbes versus tons of microbes—these studies effectively demonstrate that the microbiota may have meaningful effects on behavior and memory, even if less extreme than the differences exhibited by the extreme conditions of this experiment.

We can speculate that the microbiota improves the survival odds of the host it inhabits by making it more cautious or by improving its memory. It may be that modern humans are the result of generations of microbiotas helping our ancestors make wise, life-extending decisions. While the role they play in our personality and intelligence is still unclear, gut bacteria are certainly doing more than just helping us digest food. And although you may not be able to blame your recent course of antibiotics for forgetting your wedding anniversary, when listening to your gut, it may be the biochemical whispers of microbes that you perceive. While the microbiota is confined to the digestive tract, it is clear that its influence stretches beyond these confines. Chemicals produced by these bacteria can penetrate the walls of the intestine and seep into circulation and reach the brain. Researchers are actively pursuing the identity of these chemicals to discover how they can impact our mental state.

THE PERSONALITY TRANSPLANT

Microbiota transplants can confer physical characteristics from the donor to the recipient. The transplant of an "obese" microbiota into a lean mouse makes the mouse gain weight; similarly a "lean" microbiota transplant protects mice from gaining weight. But if the microbiota can influence brain function, could transplanted microbes change a person's mood or personality? Could a "happy" microbiota be used to combat depression?

In 2011 a research group at McMaster University in Ontario, Canada, set out to address whether gut microbes can transfer personality types the way they can transfer body types. The scientists were working with two different strains of laboratory mice. One, called Balb/c, were more anxious, the Woody Allens of lab mice. The other, called NIH Swiss, were like the Italian actor/director Roberto Benigni, more gregarious and extroverted in nature. To assess how nervous or extroverted these mice were, the scientists placed them on an elevated platform and recorded how long it took them to step down. A long delay before stepping off the platform indicated that the mice were feeling nervous about their precarious situation. The more confident the mouse, the quicker it would leap off. The Woody Allen mice spent an average of four and a half minutes carefully making their way off the platform. The Roberto Benigni mice hopped down within seconds.

Next, the scientists switched the microbiota of the two strains of mice and repeated the platform test. When the Woody Allen microbiota was transplanted into the Roberto Benigni mice, these previously self-assured mice now took over a minute to get off the platform. A transplant of the Roberto Benigni microbiota into the previously apprehensive Woody Allen mice reduced their step-off time by over a minute. By switching the microbiota from these two groups of mice, the level of anxiety and subsequent behavior of the mice could be significantly altered—behavior depended upon which microbes were living in their gut.

The researchers found that the microbiota transplants affected the levels of brain-derived neurotropic factor (BDNF) in the hippocampus. BDNF is a protein whose function has been linked to diseases such as depression, schizophrenia, and obsessive-compulsive disorder. Low levels of BDNF in the hippocampus are associated with anxiety and depressive types of behavior. After receiving the Woody Allen microbiota, the formerly outgoing Roberto Benigni mice not only became more fearful, they also had measurable changes in their brain chemistry.

From a scientific standpoint, it is unclear how the behavior change happens. Somehow the microbiota affects the levels of BDNF (and potentially other chemical messengers) in the brain. These chemical changes accompany changes in the mood and behavior of the host organism, in this case the mouse. How can bacteria at the end of the digestive tract change the expression of a protein at the top of your skull? We have known for a long time that the brain is physically and chemically wired to the gut. This wiring informs our brain when we are hungry, and it makes sense that any organism with a gut and a brain would connect the two, given that food is critical to survival. But it is becoming evident that messages from the gut are a lot more nuanced than just "feed me."

AN UNSUPERVISED DRUG FACTORY

As the bacteria in the microbiota consume MACs they manufacture, in addition to SCFAs, a huge assortment of molecules. Some of these molecules end up in our circulation and are disseminated throughout our bodies. Many of these molecules can be toxic and are cleared by our kidneys and excreted in urine. (Patients with kidney failure must undergo regular dialysis to rid themselves of these microbiota-produced chemicals.) Some microbiota chemical products resemble drugs and actually replicate the design of our body's own chemical messengers. Many of these molecules can be absorbed through the intestine and interact with our intestinal neurons, immune cells that reside in intestinal tissue, or can be absorbed into our bloodstream and circulate to the brain. These bioactive chemicals, produced by gut bacteria, bathe our own cells, pass signals to our neurons, and potentially affect our minds. Our microbiota is a drug factory dispensing pharmaceuticals from our gut—with direct access to our brain.

Why the microbiota produces drug-like chemicals is unknown. Perhaps through the action of some of these chemicals our appetite is increased, resulting in more nourishment for our gut bacteria. Perhaps these chemicals serve another as yet undiscovered function within the gut that benefits the bacteria that produce them, such as modulating gut motility or impacting immune function. More studies are needed to understand what these chemicals are doing and to gain a better sense of the trillions of pharmacists that are dispensing them.

We have to remember that these bacteria are not consciously aware and are not designing molecules for the sole purpose of manipulating us. But the following hypothetical situation may help illustrate how modulation of our behavior by bacteria could be reinforced and then become an ingrained facet of our biology.

Imagine a bacterial species that is exceptional at consuming pectin, a polysaccharide that is abundant in many types of fruit, like citrus. While this bacterium is consuming MACs from the citrus you ate, it might experience a mutation in its DNA. These biological mistakes in DNA replication happen quite frequently in bacteria, and in most instances, spell trouble for the microbial owner of the DNA, typically leading to its death. In rare instances, however, such mutations may lead to the generation of an interesting, new molecule. If, by chance, one of the billions of pectin-consuming bacteria makes a new molecule that happens to stimulate your desire to eat citrus, that bacteria just stumbled upon a means to modify your behavior in a way that will benefit it and its progeny.

An important part of this story is that such a series of events is extremely unlikely. Chances are small that a bacterium will make just the right chemical to change your craving for citrus, and that the same bacterium will profit when fruit is eaten (i.e., also be a good pectin utilizer). But considering the eons over which we have coevolved with microbes, the trillions of microbial cells within each of us that replicate their DNA every thirty to forty minutes, and the billions of humans on the planet, odds are that a microbe will hit the jackpot occasionally. Once a microbe has wandered into something that gives it a competitive advantage, even if just by dumb luck, that microbe will become more abundant. Such clever (or lucky) microbes will get passed from parents to their children and the microbe's manipulation of human behavior will persist over time. These types of scenarios are currently being played out in your body.

THE MICROBIOTA'S TOXIC WASTE

Because few bacteria actually hit the jackpot, many of the chemical compounds circulating in your body are just metabolic waste of the gut microbes. Although these molecules may still profoundly influence human biology, it may not be in ways that provide any selective advantage for that bacterium. Some of these molecules, like the short-chain fatty acids, have positive health effects, while others don't.

One of the liver's many functions is to detoxify chemical waste generated by our microbes. If the liver fails, these toxic substances can cause major cognitive problems, known as hepatic encephalopathy. As these molecules amass in the blood, they cross the barrier into the brain and wreak havoc on normal neurological function. Two common treatments for hepatic encephalopathy target the microbiota to reduce the amount of microbes in the gut, and thereby reduce the amount of chemicals they produce. One treatment is lactulose, which hastens gut transit to flush out the microbes and their metabolites more quickly. Another treatment is rifaximin, an antibiotic that decimates gut microbes. Before lactulose and rifaximin were available, surgical removal of the patient's colon (and colonic microbiota) could effectively treat the microbiota-induced mental dysfunction associated with liver failure.

The kidneys, like the liver, are charged with eliminating many of these microbiota metabolites via excretion into the urine. Researchers can monitor urine as a way to track what the microbiota is doing. If the kidneys fail, blood can become laden with microbiota waste and cognitive dysfunction can ensue. Filtering these molecules out of blood using dialysis is one means of keeping them at low levels. In the future it may be possible to reprogram the microbiota, or control its function through diet, to minimize toxic waste production and the need for dialysis.

One of the best-studied toxic microbiota by-products is trimethylamine-*N*-oxide, TMAO. Researchers at the Cleveland Clinic in Ohio discovered this molecule while searching for blood-borne chemicals that can predict the onset of cardiovascular disease. Identifying molecular markers of impending health problems, such as a heart attack, can serve as a valuable warning system for people and provide insight into the genesis of disease. The group at the Cleveland Clinic compared the chemicals found in the blood of people undergoing cardiac evaluation. They found that if TMAO was abundant in a person's blood it could be a predictor of impending heart attacks and strokes, as well as promote life-threatening clogging of arteries. Where does TMAO come from and what can a person do to keep TMAO levels low?

Well, as you may have guessed, the microbiota is critical to TMAO production. But, consistent with what we know about risk factors for heart disease, diet also plays a major role. Red meat and other fatty foods provide the microbiota with the resources required for TMAO synthesis, specifically a type of fat called phosphatidylcholine, known more commonly as lecithin, and carnitine, a component of meat.

In follow-up studies, the researchers found that some individuals had a microbiota that did not produce much TMA, or trimethylamine, the precursor of TMAO; had lower circulating TMAO; and had a decreased risk of heart disease. Not surprisingly, dietary habits were a major factor in determining how much TMAO appeared after eating red meat; vegans and vegetarians produced much less TMAO than their meat-eating counterparts. One of the most amazing aspects of this study was that researchers found a five-year vegan willing to eat a steak for science. This longtime meat abstainer had very low serum TMAO levels after the steak dinner, indicating that the vegan's microbiota contained a collection of bacteria that were not adept at making TMA. While the researchers couldn't (or didn't) try to convince the vegan to continue to eat meat, they did perform an analogous study in mice to see if regular meat consumption could change the microbiota to one that does produce a lot of TMA. Mice with a microbiota that produced small amounts of TMA, when placed on a steady diet of carnitine, eventually became high-TMA producers. The increase in TMA was accompanied by a change in the composition of their microbiota to one that presumably contains more TMA-producing bacterial species.

This study helped to establish how the overconsumption of red meat could lead to heart disease. The microbiota, through its production of TMA from carnitine, can seriously impact the health of its host. This study also reinforces the profound impact that diet can have on two facets of the microbiota: the species of bacteria that make up the community and the chemical reactions this collection can perform. Say, for example, two people, one who eats a predominantly plant-based diet, the other an omnivore who frequently eats meat, sit down for a steak dinner. You might think that by eating the same exact meal the chemical reactions occurring in both guts would be similar, if not identical. But the microbiota that rarely encounters meat might look more like that from the vegan in the Cleveland Clinic study and produce very little TMA. In contrast, the omnivore's steak dinner likely will result in more microbiota-produced TMA. Same meal, different chemical consequences. Now if the individual on the plant-based diet decides to eat meat more frequently, the microbiota will respond to this change. Two months later the same dinner may result in a similar amount of TMA production between the two individuals.

On the one hand, diet is important from the standpoint of providing the starter material that the microbiota acts upon—if you eat less meat, even if you have a microbiota that efficiently produces TMA, without the input molecules less TMA will be made. However, if you eat less meat for an extended period of time, your microbiota will likely become less capable of producing TMA, minimizing the TMAO in your blood should you partake in an occasional steak. The fact that each person harbors a unique microbiota capable of producing different types and amounts of bioactive molecules, combined with the dietary influence over their production, speaks to the need for the development of new technology. These tools would allow each of us to monitor aspects of microbiota function that are critical to our health.

Decades from now, TMAO may stand out as one of the most important aspects of microbiota function that is routinely monitored. Or, perhaps more likely, it will be just one of hundreds considered in the functional profile of an individual's microbiota. From what we know now, TMAO appears to do nothing to our biology that enhances the success of the microbiota members that are involved in its synthesis. But it provides a very good example of how microbial metabolism in our gut is producing strange and novel compounds that can have very real health consequences.

THE BRAIN'S TWO-WAY CONFERENCE CALL WITH TRILLIONS

Communication between the brain and microbes in the gut is bidirectional. Not only can the microbiota affect things like mood and memory, but the brain can also have a say in which microbes live in the gut. If you induce stress or depression in a lab animal by removing the animal from its mother, the composition of its microbiota changes. How, exactly, this happens, no one knows for sure. Perhaps the body's fight-or-flight response is to blame. When an animal perceives a threat from a potential aggressor its body releases an assortment of hormones and neurotransmitters that prepare it to attack the predator or flee from harm. The body's responses to a fight-or-flight situation include increased heart rate, release of energy stores to fuel muscle, increased blood flow, and changes in gastrointestinal motility. When digestion slows or stops in response to threat, our microbes are keenly aware of the change in environmental conditions within the gut. Microbes that are best adapted to the new, slow-moving pace of food will become more plentiful and the microbes that rely on quick intestinal transit times will become less abundant, thereby changing the composition of the microbiota.

Stress, the microbiota, and the immune system are interwoven in a complex set of interactions. Maternal-separation-induced stress in lab animals can cause lasting changes in the microbiota that are durable into adulthood. It may be that long-term effects on the immune system from stress can lead to persistent changes in the microbiota, even after the stressful event has passed. Or maybe microbiota disruption from stress can cause lasting immune system changes, which can then cycle back into further microbiota modification. Infant rhesus monkeys that are separated from their mothers not only have a different microbiota composition than before the separation, they are also more susceptible to opportunistic infection. If the immune response is not tuned properly, it can cause further deterioration of the microbiota, setting off a spiral of decline.

Mice infected with intestinal pathogens are more anxious than noninfected mice. (Another case of microbes affecting behavior.) If that anxiety leads to microbiota changes that allow for a more robust or long-lasting pathogenic infection, intestinal inflammation can worsen. Inflammation in the gut negatively impacts the composition of the microbiota, providing another example of a negative spiral. Similarly, when an anxiety response is accompanied by intestinal motility changes such as diarrhea or constipation, the balance in the gut can be shifted to one that further favors the pathogen. It is possible that individuals experiencing functional bowel disorders like irritable bowel syndrome (IBS) and motility disorders, and even those suffering from inflammatory bowel disease (IBD) are victims of such an imbalance. This scenario exemplifies the negative consequences of a malfunctioning brain-gut axis. Because chemicals secreted by the microbiota can affect mood and mood can affect the microbiota, identifying what initiates the cascade of events is difficult. Both IBD and IBS are marked not only by gastrointestinal symptoms such as chronic diarrhea, constipation, and/or bloating, but also by mood disturbances such as depression, anxiety, and increased pain perception.

So what comes first? Did a stressful episode cause a deleterious microbiota change? Or did the microbiota disturbance occur first, leading to oppressive anxiety or depression? Understanding and treating these types of diseases is incredibly complex because they involve a malfunction in the communication between our most complex ecosystem, the microbiota, and our most complex organ, the brain.

Some see beneficial bacteria as a possible escape hatch for those caught up in a crippled brain-gut axis. A class of probiotic bacteria called psychobiotics aims to improve psychiatric symptoms by delivering psychoactive compounds from the gut to the brain. By adding bacteria to the gut that can synthesize chemicals to normalize behavior, it is possible that a healthier brain-gut connection can be rebuilt. There is growing evidence that supplementing the gut with probiotic bacteria improves behavior in animal models of stress and depression. Preliminary studies in humans also show promise in easing symptoms associated with chronic fatigue syndrome and IBS with probiotics. Even healthy volunteers who consumed a cocktail of two types of probiotic bacteria daily for thirty days described feeling less anxiety and depression after the probiotic therapy. Despite cause for optimism, it is important to point out that these are preliminary studies and that more placebo-controlled trials are required to determine how probiotic bacteria can best be used to treat diseases like IBS and IBD and mood disorders such as depression and severe anxiety. Therapies may need to be personalized. But these studies are a reminder that our internal microbes are playing a role in diseases that affect both the brain and the gut.

CHEMICAL SPILLS OUT OF THE GUT

The incidence of autism spectrum disorders (ASD) is reaching epidemic proportions. According to the Centers for Disease Control, one in sixty-eight children is affected by ASD, a rate that has steadily increased over the past decade. There have been a number of risk factors identified in the onset of ASD, including parental age and profession and certain genetic factors. But the ever-growing list of possible causes for ASD, some still under investigation and some completely debunked, demonstrates the difficulty in determining the mysterious etiology of this disorder. The gut microbiota first made its way onto the list of possible ASD risk factors after medical professionals noted that many children with ASD also suffer from gastrointestinal issues such as chronic diarrhea, constipation, gastrointestinal cramping and bloating, and even more serious conditions such as IBD.

Much research has gone into cataloging the differences in the microbiota composition between children with ASD and those without it. But attempts at compiling a list of "bad" bacteria that are overly represented in ASD children and "good" bacteria that are lacking has been as frustrating an endeavor as generating a list of causes for this disease. While marked differences have been reported in the microbiota of children with ASD, many of the studies are contradictory. Perhaps the lack of a microbiota signature for ASD should not come as a surprise considering the individuality of each person's microbiota and the range of severity and multitude of subtypes observed in ASD. It is not a stretch to imagine that microbiota disturbances in those with ASD would manifest in different ways. While studies were unable to pinpoint a reproducible microbiota abnormality in ASD children, they do demonstrate that children with ASD typically have a gut microbiota that deviates from what is considered normal. But are these differences meaningful to the etiology and progression of ASD or are they just unrelated side effects of the disorder? Could the disorder be treated or prevented by reprogramming the microbiota?

In 2013, a group of scientists from Caltech led by Sarkis Mazmanian made major progress toward a better understanding of the relationship between the microbes in the gut and ASD. These scientists were studying a group of mice that were born to mothers whose immune systems had been activated as if they had been exposed to an infection. For a subset of human ASD patients, an extreme immune response of the mother to an infection during pregnancy appears to contribute to ASD onset. Mouse pups born to mothers with a chemically induced immune response display many of the gastrointestinal and behavioral hallmarks of human sufferers of ASD. These pups have greater permeability in their intestine, meaning the "grout" that connects the "tiles" of intestinal cells is incomplete, increasing the leakage to small chemical molecules produced by the microbiota. These mice are more anxious, engage in repetitive behaviors, and don't communicate and socialize like normal mice. And as with many human sufferers of ASD, the microbiota of ASD mice looks abnormal.

The Caltech research team wondered if they could affect the ASD-like symptoms displayed by these mice by introducing beneficial bacteria. The researchers gave the ASD mice the very common human gut bacterium *Bacteroides fragilis. B. fragilis* repairs intestinal leakiness by promoting the colonic epithelial cells to secrete their own "spackle," molecules that patch up the leaky grout in the intestinal lining of mice. They reasoned that if the leakiness could be repaired, fewer chemicals would escape the confines of the gut, thereby reducing the severity of ASD symptoms. Their bet paid off.

Supplementing the ASD mouse gut with *B. fragilis* corrected intestinal permeability and even brought the microbiota composition closer to that of the normal mice, though it still remained notably different. More surprisingly, the *B. fragilis* treatment also ameliorated many behavioral issues with the ASD mice. Treated mice were less anxious, reduced their repetitive behaviors, and displayed improved communication. Although the socializing defect remained, the *B. fragilis*–induced ASD improvement was striking.

Before you rush out to buy stock in *B. fragilis* supplements, there are two things you should know. First, *B. fragilis* is not commercially available because, like most other prominent bacteria from the gut, it would require human studies before being made available to the public. Second, the researchers found that the improvement in ASD observed after consuming *B. fragilis* was not specific to this species of bacteria. Another related human gut resident, *Bacteroides thetaiotaomicron*, also eased ASD symptoms, opening the possibility that several bacterial types may be capable of producing a similar outcome. Perhaps the most potent microbe or microbes will be specific to the type of ASD, to the microbiota of the individuals being treated, or to their genetic makeup. Clinical trials, currently under way, are required to identify safe and effective bacterial strains for humans. If the addition of microbes can help people suffering from ASD, it is possible that the use of many friendly strains of bacteria, a microbiota-therapeutic equivalent of bet-hedging would provide a broadly effective treatment.

The researchers identified specific chemicals produced by the microbiota circulating in the ASD-like mice. One of these molecules, abbreviated EPS, was forty times more abundant in the bloodstream of the ASD-like mice compared to that of normal mice. As *B. fragilis* treatment repaired the leakiness of the gut, it reduced EPS levels to normal. So can EPS alone cause behavior that looks like anxiety? To test this possibility, the researchers injected EPS into healthy mice and found that it induced ASD-like behavioral changes. Now this doesn't mean that EPS is the only chemical, or the most important chemical related to human autism; remember, these studies were performed in mice. But clearly the microbiota is capable of synthesizing specific chemicals within the gut that can affect behavior. If the gut is more porous than it should be, too much of these microbiota-made chemicals can escape the boundaries of the intestine and enter the bloodstream.

The unsupervised drug factory in our gut produces a range of chemicals about which we have very little understanding. Some of these chemicals, if leaked into the bloodstream at high concentration, could cause abnormal behaviors or moods. Adding beneficial bacteria, in this case *B. fragilis*, patched the intestinal leakiness in the ASD mice, and along with the repair came a decrease in circulating chemicals produced by the microbiota. In fact, the concentration of more than a hundred different bacterial chemicals in the bloodstream of the mice changed after *B. fragilis* treatment, some of them decreasing to levels found in normal mice. Whether a disruption of the microbiota is sufficient to cause certain subtypes of ASD is still under investigation. However, the potential role of the microbiota in contributing to ASD appears to have some promising leads.

The connection between ASD and the microbiota is just one example of the ties that connect the microbiota to a variety of behavioral disorders including schizophrenia, obsessive-compulsive disorder, and depression. Microbes in the gut, through their ability to synthesize neurologically active molecules, have influence over aspects of our biology that on the surface seem unconnected with the gut. That gut feeling you have may in fact be a chemical message sent to your brain by one of your internal microbial inhabitants. And depending on which types of bacteria reside in the gut, these messages may combine with a person's genetic predisposition to increase or decrease the chance that a behavior disorder manifests. Research into how the microbiota feeds into the brain-gut axis offers hope that behavioral diseases might be managed in the future by rationally manipulating the microbiota. Present-day medicine is not able to change the microbiota in a predictable manner. But we do know that there are many powerful levers to change the inhabitants of the gut, including diet and exposure to environmental microbes. These factors may provide avenues to affect the brain through its connection with the gut.

FERMENTED FOODS ENTER THE CONVERSATION

Because much of the work related to the microbiota-brain axis has been performed in laboratory animals, we need to be cautious in interpreting the findings. Animal studies demonstrate unequivocally that there is a meaningful connection between the bacteria in the gut and the brain, a connection that is most certainly conserved in humans. However, trying to extrapolate specific behaviors or behavioral changes observed in connection with certain gut microbes in mice to humans would be ill advised. The human brain and microbiota are distinct from those of the mouse. Studies on humans are needed before we can understand the specifics of how the human microbiota affects ASD, depression, and anxiety disorders, as well as human personality and mood.

In 2013, a group of scientists at UCLA set out to determine whether the human brain might also be influenced by bacteria in the gut. Twelve women, all of whom were free of gastrointestinal and psychiatric symptoms, consumed a yogurt containing four different types of bacteria twice a day for four weeks. For comparison, two other groups of women either consumed a twice-daily placebo (a bacteria-free yogurt) or had no intervention at all. The study was double-blinded, meaning neither the participants nor those conducting the study would know who was eating bacteria and who wasn't until after the study was completed. Using functional magnetic resonance imaging (fMRI), they scanned the brain of each participant before the intervention and again after the four-week trial. All women underwent fMRI scans while resting and also while performing a task in which they matched faces displaying negative emotions such as fear or anger. This matching activity was chosen because people suffering from certain types of anxiety disorders show an altered brain activity pattern by fMRI when performing this mental exercise.

During both the resting scans and those performed during the face-matching test, the scientists saw differences in brain activity between the women who had consumed the bacteria-containing yogurt and those who had not. The changes occurred in several regions of the brain involved in processing sensory input and emotion including the frontal, prefrontal, and temporal cortices and the periaqueductal gray area. The affected areas are important in anxiety disorders, pain perception, and irritable bowel syndrome. It may seem almost impossible to believe that just four milk-fermenting bacterial species swimming amongst the hundreds of species already living in your gut can have such an extensive impact, affecting multiple regions of the brain.

But as this study showed, eating two yogurts a day for one month is enough to change the pattern of your brain's activity in a measurable way. By demonstrating a gut bacteria–brain connection in humans, this scientific breakthrough has unearthed numerous follow-up questions. What do the differences in brain scans mean with respect to the mental health of the probiotic consumer? How are these four probiotic bacteria able to affect brain function? Is it through chemicals they secrete or is it less direct? Do most types of bacteria affect brain function, or is this ability limited to a select few species? Can microbes be used as effective treatments for mental illnesses with minimal worry over side effects that plague so many of the drugs currently used to treat these types of diseases? These and other related questions will be an active area of research in the coming decade.

Whether pathogenic bacteria can induce anxiety or probiotic bacteria can help ease depression in humans, as has been reported in mice, remains to be seen. If and how bacteria can be used to treat ASD in humans is also something future studies will answer. Clearly there is a lot more work that needs to be done before we can understand how the bacteria in our gut impacts our brain and how we can ensure that this relationship is beneficial to our mental health. All these preliminary studies lay the groundwork for a better understanding of the role the microbiota plays in our mental health and how we can manipulate this community for optimal health and mental function. Deciphering the complexity of the brain presents a monumental scientific challenge. Throw trillions of bacteria into the mix and it's clear that unraveling this brain-microbiota connection will take some time. But while more research in this area is needed, we now have an understanding that our brain and microbes are engaging in an ongoing dialogue, one that has huge implications for our mental well-being.

BUILDING LIFELONG BRAIN-GUT-MICROBIOTA ALLIES

When human babies are born, the tissues in their bodies are extremely immature, requiring years of development. The newborn's infant gut is leaky, its immune system is naive, and the infant brain requires years to forge necessary connections. Because babies are born without a microbiota, the brain-gut axis forms during the time period in which the gut microbiota is becoming established. How do differences in microbiota assembly affect the brain-gut axis? We know that the first year of a child's life is marked by a seemingly chaotic assembly of the microbiota. Do the specifics of early microbiota assembly impact how the brain-gut axis is formed and determine how it will function throughout the life of the child? If something goes wrong during microbiota assembly, can it be repaired in order to establish a healthy brain-gut connection? Does the final, stable composition of a child's microbiota impact how his brain will function throughout his adult life?

The human brain is a work in progress throughout life, but the first few years of life are exceptionally critical. Our experiences in childhood have a lasting impact on the physical structure of our brain and our mental well-being, including our risk for depression, anxiety, and other mental disorders. Because the beginning of life represents such a crucial window of time for development of the brain and the microbiota, perhaps these two phenomena are intimately intertwined. The infant brain is likely to be more susceptible to the effects of microbe-produced molecules circulating through the blood than the brain of an adult. As the child experiences dietary milestones such as making the transition from breast milk to solid food, possibly eating meat for the first time, and consuming his first fermented foods, the resulting changes in microbiota composition will translate into changes in the types of microbiota-produced chemicals coursing through his body. Other significant events, such as a child's first intestinal infection and his first round of antibiotics, can also steer microbiota-host relations in a new direction, for better or for worse. The unsupervised drug factory begins dispensing at birth and changes the types and amounts of compounds it doles out as the child's brain is developing.

Our current understanding of how gut microbes influence the development of the human infant brain is, like the infant's microbiota, undeveloped. Microbiota-free mice have disturbances in their perception of pain and levels of anxiety, both of which can be repaired upon introduction of gut microbes. However, this microbiota introduction needs to occur early in life or else the disturbances last into adulthood. How microbial exposure early in life influences the function of a child's brain requires formal testing in humans. As adults, our response to stress, our ability to learn and remember, and even the finer points of our personality could be a result of the state of our microbiota early in life.

Perhaps individuals with paralyzing fear or irrational phobias or those who participate in risky behaviors like extreme sports can partially attribute those behaviors to their gut microbes. There are a handful of central nervous system diseases, such as ASD, hepatic encephalopathy, and multiple sclerosis, in which a change in microbiota composition accompanies changes in symptoms. Maybe the changes in our microbiota brought about by our constant use of antibiotics and our low-MAC diet, in addition to contributing to obesity and heart disease, are also a factor in the rise in ASD, depression, and anxiety disorders in the West. Maybe if we can repair the damage to our microbiota through diet, increase consumption of probiotic bacteria, and constrain the use of antibiotics and antibacterial soaps and cleaners we could improve our mental health. But these are all big maybes at this point.

It would be premature to make any specific, scientifically based recommendations for ways to improve the health of your brain-gut-microbiota axis. But while this avenue of exploration is just opening, its potential to be helpful in the future is great. Even in the absence of data from placebo-controlled clinical trials, it seems not too much of a stretch to assume that improving the overall health of your microbiota could have a positive impact on your mental well-being. Eating a diet replete in MACs to feed this community, limiting antibiotic use, breast-feeding, and safely increasing environmental exposure to microbes all have the potential to improve the state of the microbiota and perhaps improve the state of the brain. Dr. Thomas Insel, director of the National Institute of Mental Health, sees the incredible potential of the microbiota to affect how we approach the treatment of mental disease: "How these differences in our microbial world influence the development of brain and behavior will be one of the great frontiers of clinical neuroscience in the next decade."

The connection between our microbiota and our brain beautifully illustrates the far-reaching implications our microbes have for all aspects of our health. As our understanding of the microbiota improves, we are increasingly realizing that every aspect of human biology is touched by our associated microbes in a direct or an indirect way. It's evident that we need to stop thinking about our various organs and their associated diseases in a reductionist way; it would have been hard to imagine, even just a few years ago, that disorders of the brain could have their roots in the gut. The fact is that our body is a complex ecosystem and that all its parts are interconnected. The disruption of one aspect of the microbiota will cascade throughout the entire body. Another more positive way to think about it is that by strengthening a single part of our ecosystem, we can support our overall health.

CHAPTER 7

Eat Sh!t and Live

CHANGING YOUR MICROBIAL IDENTITY

Your human genome determines a lot about you, and there is not a lot you can do to change your DNA. Some people are destined for disease solely because of the genes they inherit. Even in cases where the genetic flaw is known, changing the genetic material in an individual's genome to treat or prevent disease, known as gene therapy, is extremely difficult.

Unlike the unyielding human genome, the gut microbiota offers more flexibility and is an effective avenue to improve health or treat disease. These microbes connect to most aspects of our health, and in some cases cause disease, but they are much more malleable than the human genome. Just think of the possibilities. If a pathogen within your gut is secreting a toxin that is making you sick, there is hope of eradicating this nasty character from the ecosystem, along with its ill effects. Alternatively, if you (or your doctor) were to discover that your microbiota happens to be missing an important function or a key bacterial species, adding a new microbial member could ameliorate the deficiency. This plasticity or ability to change the gut microbiota, also known as "reprogramming" the microbiota, has a promising future, and is already proving to be a powerful way to improve human health.

PARTY CRASHERS

Gastroenteritis, also known as stomach flu, food poisoning, traveler's diarrhea, or Montezuma's revenge, is something most of us have had firsthand experience with at some point in our life. Infectious diarrhea is one of the most common childhood illnesses worldwide and is a leading cause of death in the developing world for children under the age of five. While mortality from gastroenteritis is incredibly low in the Western world, we are in no way impervious to its effects. In the United States, infectious diarrhea results in the hospitalization of more than one million people each year with many more millions treated in outpatient services. Collectively, Americans experience approximately 200 million episodes of acute diarrhea per year, second only to the common cold in prevalence. There are a number of microbial culprits behind these episodes, such as norovirus of cruise ship infamy; *Salmonella* bacteria, which can lurk in everything from undercooked eggs to jars of peanut butter; or parasites like *Giardia* that often spread through contaminated water. Because these infectious microbes inhabit our environment, it is not that difficult to come into contact with one occasionally and then suffer the consequences. Children, the elderly, and immunocompromised individuals are especially prone to infectious diarrhea and spend more time in day care centers, schools, senior care facilities, and hospitals, where contagious microbes can spread rapidly.

When you ingest a microbe that is capable of causing disease, a number of factors can determine whether illness will result. Many bacterial pathogens travel through the digestive tract to the large intestine where they are confronted with the bustling, bacteria-filled microbiota. These beneficial microbes serve as competition to pathogenic invaders such as *Salmonella* and *Clostridium difficile*, or *C. difficile*. You can think of these pathogens as party crashers—uninvited, unwelcome guests.

"Colonization resistance" is a term used by scientists to describe the protection the microbiota provides against invading pathogens. This protective effect is both direct and indirect. First, the microbiota can take up physical space and precious resources, making it difficult for pathogens to find both room to expand and food to fuel their growth. Second, some gut microbes can unleash chemical warfare in the form of bacteriocidal chemicals that kill pathogens. Finally, indirectly, the microbiota can coax the immune system to beef up its defense to help fight the infection. With all of the positive things that the microbiota does to help us fend off these assailants, it is not surprising that antibiotics, which decimate the microbiota, provide pathogenic bacteria a window of opportunity to take hold.

FIGHTING FIRE WITH FIRE

Antibiotic use is one of the biggest risk factors for infection by *C. difficile*, a bacterial pathogen that can cause severe diarrhea and intestinal inflammation. Taking a round of antibiotics is like setting our microbial ecosystems ablaze. As in the aftermath of a forest fire, close inspection will show that while a few things survive, overall the landscape changes dramatically. After the fire, new seedlings that might not previously have had the space or resources to grow can now take hold. Some of these fledgling plants may be productive and healthy members of the rebuilding ecosystem, like mutualistic bacteria, but others may be invasive and detrimental, like pathogenic bacteria. Over time, the forest matures and, optimally, regains stability in a way that favors a balance of good plants that can coexist in harmony. However, occasionally an invasive weed can take hold and change the landscape of the forest indefinitely. Such is the story of *C. difficile* associated disease, or CDAD.

CDAD is responsible for the death of approximately fourteen thousand Americans each year; currently another ten times that number are battling *C. difficile* infections. Those whose infections are especially refractory to multiple rounds of antibiotics have a coin's-toss chance of surviving. *C. difficile* is most notoriously found in hospitals, but it can also lurk in swimming pools, raw vegetables, and pets. If you think that pouring gallons of chlorine into your pool, scrubbing your vegetables with antibacterial soap, and loading up your pets with antibiotics will keep you from being exposed to *C. difficile*, think again. An estimated 2 percent to 5 percent of people are unknowingly carrying *C. difficile* around in their microbiota. In a hospital setting that number rises to 20 percent, and half of the residents of long-term-care facilities are *C. difficile* carriers. Just because you never have suffered from CDAD doesn't mean that your gut is free from *C. difficile*. In fact, for most carriers, *C. difficile* is a perfectly well-behaved member of the microbiota and may never cause disease. But if something disrupts the microbiota—a round of antibiotics, for example—the previously well-mannered *C. difficile* may take advantage of the disturbance and multiply to high numbers to cause problems.

Once it starts wreaking havoc in the gut, *C. difficile* can cause life-threatening diarrhea and intestinal inflammation and be extremely difficult to eradicate. Until recently, the treatment for recurrent *C. difficile* infection was more antibiotics—the equivalent of lighting a second forest fire—in the hopes that good bacterial species would repopulate. One problem with this strategy is that *C. difficile* can wait out the fire in a highly antibiotic-resistant state of suspended animation called a spore. Bacteria that form spores are especially hard to eliminate because spores can survive conditions not usually suitable for life such as boiling, desiccation, subzero temperatures, and even the vacuum of space.

After the antibiotic assault is over, these stealthy spores can reemerge. However, in a freshly cleared gut landscape, one that is full of open space and unused resources, the germinating spores can further expand. Some patients suffering from CDAD have a microbiota that is largely composed of *C. difficile*. In these individuals, the pathogen has caused the mass extinction of hundreds of bacterial species and co-opted the gut completely for itself. *C. difficile* contains multiple genes that produce toxins. When *C. difficile* is in low numbers within the microbiota, it refrains from harming the intestine. However, once *C. difficile* becomes abundant, it can unleash its toxins, causing both damage to the intestinal wall and painful diarrhea.

If you happened to develop a *C. difficile* infection that could not be eliminated by antibiotics, until recently there were not many options for treatment. Failure of one round of antibiotics would be followed by more and different antibiotics in a series of Hail Mary attempts to knock down *C. difficile* and give beneficial bacteria the chance to reestablish. If continued antibiotic therapy didn't work, doctors were left with little other choice but to surgically remove the infection and diseased intestinal tissue. While this approach can be successful in eliminating CDAD, it is a last resort that has lifelong consequences even for patients with the best outcomes. But what if instead of indiscriminately torching the microbes in the gut, or surgically removing them, you could deliberately introduce healthy microbes back in? Could restoring a community of good gut microbes effectively limit resources to *C. difficile* and smother the infection?

I HAVE TO DO WHAT?!

In 2013, a group of scientists and physicians at the Academic Medical Center in Amsterdam set out to test the idea that an infusion of beneficial bacteria could stop the vicious cycle of recurrent *C. difficile* infection. They performed a randomized, controlled clinical trial in which individuals with recurrent *C. difficile* infection were treated either with antibiotics alone or antibiotics followed by a fecal microbiota transplant (FMT), also known as bacteriotherapy or stool transplant. For those who have not heard of this procedure, the name provides a quite accurate description. In an FMT, stool is taken from a donor and introduced into the recipient's intestine. An FMT can be performed from the top down, through a tube running through the nose into the gut, or bottom up, administered rectally into the colon by an enema or colonoscope. For both procedures the fecal material is liquefied (often in a blender), then strained, and is ready to go. Before you dismiss this approach as having way too much of an ick factor, remember that people with CDAD are dealing with a life-threatening disease. Under such circumstances, most of us would be more than willing to endure a little grossness to regain our health.

To qualify for the Dutch study, all participants must have already tried antibiotic therapy that was unsuccessful. After a single FMT an astounding 81 percent of these recurrent infections were cured, compared to the 31 percent cure rate in the group that tried another round of antibiotics alone. A second FMT was performed on the remaining 19 percent of nonresponders and the overall cure rate climbed to 94 percent. This cure rate was so high that the researchers felt it was unethical to continue the study and abruptly terminated it and offered FMTs to all the participants. The overwhelming success of this crude microbiota reseeding approach has made using FMTs a much more palatable option.

While documenting the success of FMTs in a randomized trial was an important step in the procedure's gaining widespread recognition, FMTs were in fact used in the United States more than fifty years ago. In 1958, Dr. Ben Eiseman, chief of surgery at Denver General Hospital, published the first report that a "fecal enema" could cure pseudomembraneous colitis. It wasn't until almost twenty years later that *C. difficile* was identified as the causative agent for pseudomembraneous colitis. Dr. Eiseman and his colleagues in Denver didn't really understand what was at the root of this debilitating disease. But they presumed that somehow the "balance of nature" within the gut of these patients had been lost and could be regained with a transplanted community. Veterinarians have been treating animals with FMTs for more than a hundred years, sometimes even transplanting feces from one type of animal into another, a procedure called transfaunation. But the use of feces as medicine goes even further back than that. From fourth-century China there are records of severe diarrhea being treated with a tonic of fecal material called "yellow tea."

The field was abuzz with excitement after the 2013 FMT study was published. A transplanted microbiota as therapy opened many intriguing possibilities about the future of microbiota-based treatments. FMTs are a perfect demonstration of how a disease can be ameliorated by simply repairing the microbiota. The medical community is currently examining the extent to which illnesses can be ameliorated or even cured through repairing the microbiota. More than forty clinical trials are under way to explore the efficacy of fecal transplants to treat a variety of diseases, including inflammatory bowel disease and obesity. But can we expect FMT's success to extend to diseases beyond CDAD? And why does FMT work so well to eradicate *C. difficile*? To understand these questions, we first have to explore what is happening when antibiotics compromise colonization resistance and *C. difficile* takes control of the gut.

ANTIBIOTICS—THE INDISCRIMINATE KILLER

"Antibiotic" literally means "against life." While sounding ominous, these medicines usually do take out the bad guys—the bacteria that make us sick. Most of us routinely use antibiotics and give them to our children without a second thought. We view these medicines as lifesavers, and rightly so. However, recent research shows that antibiotics, as their name implies, have a broader impact on our physiology than previously appreciated. They impact our health by crippling our resident beneficial microbes.

Humans have been using antibiotics for thousands of years. Even the ancient Greeks harnessed their antiseptic properties by applying a mash of moldy bread to wounds to protect them from infection. From one type of mold, *Penicillium*, came medicine's most famous antibiotic, penicillin. With the ability to cure previously life-threatening illnesses, antibiotics can arguably be called the single greatest advance in medicine. The effectiveness of antibiotics and their relatively few acute side effects have driven drug companies to develop and produce many different antibiotics designed to treat a variety of infectious diseases. But the development of new antibiotics can be costly, so drug companies bias their research and development efforts toward broad-spectrum drugs—antibiotics that kill a variety of different microbes. In this way, an antibiotic can be prescribed to eradicate an array of bacteria that cause everything from ear to urinary tract infections. Today, Americans represent one of the largest populations of antibiotic users in the world. In 2010, doctors handed out no fewer than 258 million courses of antibiotics, roughly eight and a half prescriptions per ten people in the United States. The rise of antibiotic-resistant superbugs is one well-documented adverse effect of our widespread use of antibiotics. But perhaps more important—and far less publicized—is the impact of these drugs on our resident microbes.

The vast majority of antibiotics are taken orally, regardless of where pathogenic bacteria are causing the problem. At first glance, this may make sense. For example after swallowing an antibiotic, some will be absorbed into the bloodstream and eventually circulate to the ear and kill the earache-causing bug. But this bodywide distribution of the drug puts all bacteria in and on your body in the line of fire. The oral route in particular puts your gut microbes directly in the drug's crosshairs. And because most antibiotics are designed to kill many different bacterial species, each dose results in significant collateral damage to the microbiota. For some individuals it can take months for the gut microbes to recover, and during that time the risk of diarrheal illness skyrockets.

David Relman and Les Dethlefsen, two of our colleagues at Stanford University, were curious about what would happen to the microbiota after multiple rounds of the potent antibiotic ciprofloxacin (trade name Cipro). This commonly prescribed broad-spectrum antibiotic is used to treat a variety of bacterial infections. Its mechanism of action is to inhibit a microbe's ability to replicate its DNA, effectively keeping bacteria from proliferating. Because Cipro is broad spectrum, it is active against most types of bacteria, both infection-causing bacteria and the friendly, mutualistic type living in the gut. David and Les wanted to determine how damaging a five-day course of Cipro would be to the microbiota, and whether it would fully recover.

Microbial abundance and diversity in the test subjects' guts plummeted rapidly upon starting antibiotics. There were ten to a hundred times less gut bacteria after Cipro treatment, and the surviving community was much less diverse than before. The microbiota was also significantly reorganized, with bacterial species that collectively made up 25 percent to 50 percent of the total organisms in the gut nearly wiped out. These results should not be that surprising, although the scale of microbiota damage was even larger than many had feared it would be. Cipro, like all broad-spectrum antibiotics, was designed with no consideration for sparing gut microbes. Despite how critical our resident microbes are for health, there is still widespread (and often relaxed) use of broad-spectrum antibiotics, partly owing to the belief that the microbiota can regrow. But is the assumption that beneficial microbes can stage a comeback and repopulate correct? Not exactly. A few weeks after the Cipro treatment, one subject's microbiota recovered to the pre-antibiotic state. The other two were not so resilient. One individual had a near complete recovery but still harbored visible antibiotic-induced damage. The third subject's microbiota was struggling to regain its pre-antibiotic composition even two months after the Cipro treatment had ended.

Many microbiotas have to endure multiple antibiotic exposures, often within a single year, so Relman and Dethlefsen tested what happens to these same individuals after a second round of Cipro. From the microbiota's perspective, the damage was even more severe. Following the second course of antibiotics, again bacterial abundance decreased, the community shifted in composition, and diversity took a hit, just as it had after the first Cipro treatment. But this time none of the individuals escaped unharmed. All three test subjects had visible, lasting Cipro-induced damage to their microbiota, even two months after the antibiotics were stopped.

None of the study participants reported any gastrointestinal symptoms, despite the massive reorganization that was occurring in their gut. Clearly, symptoms are not a reliable gauge of how much damage the microbiota sustains from antibiotics. The scientists also could not predict, a priori, whose microbiota was more susceptible to antibiotic-induced damage. So, while there is no test you can get at a doctor's office to tell you how big a blow a prescription of antibiotics will be to your microbiota, you can assume that it will be significant. And a second course of antibiotics will likely compound the disturbance to your microbiota caused by the previous course. Most of the time antibiotic treatment seems to provide a huge benefit—ameliorating infection—at no cost. But even though it may not be perceptible to you, your microbiota has sustained much damage. Much of this damage can take several weeks to repair and there are some species of bacteria that may never fully recover. Meanwhile the microbiota's ability to protect you from another invading pathogen is compromised, increasing your vulnerability to dangerous infections. While you can try to mitigate antibiotic damage by taking probiotics after treatment, the fact remains that we still don't know how to effectively return a microbiota to its exact pre-antibiotic state.

THERE IS STRENGTH IN NUMBERS

A pathogen trying to infiltrate the gut is like a foreign invader attacking an established army (the microbiota). If the invading force is small, there is little chance that it can overcome the more numerous and established resident troops. A much larger group of invaders with weapons deployed is required to mount a successful attack. But when the home force is weakened, a smaller group of invaders can prevail. Our bodies are constantly exposed to pathogenic bacteria like *Salmonella*, but because of the resistance afforded by our microbiota, it takes many pathogenic microbes to make us sick. The trillions of beneficial microbes living in the gut make it difficult for the few *Salmonella* microbes in the undercooked egg to mount a successful attack. However, a microbiota with a reduced membership, say as a result of antibiotics, is at a much higher risk of getting overcome by a smaller dose of pathogens. With this in mind, taking antibiotics looks a bit like a game of Russian roulette. In most cases, as we mentioned earlier, the damage is not immediately perceptible to you. But the blow to your microbiota could be life threatening.

Metabolic interactions between different microbial species within the microbiota create an intricate food web in which all the resources are used. When functioning perfectly, this bioreactor is a tight resource-allocation system allowing great bacterial diversity with no extraneous resources. The tightly knit food web created by the microbiota helps exclude pathogenic invaders by rapidly using up all available resources, leaving little to no leftovers for pathogens to consume. In this ideal state the internal bioreactor is stable and resistant to invasion by disease-causing microbes. However, as most of us have experienced, there are times when this resistance fails and pathogens wedge themselves into the food web, proliferating within the gut by using stolen resources for their own success.

When normal mice are exposed to *Salmonella* or *C. difficile*, no infection takes hold. However, if the mice have been treated with antibiotics prior to receiving the pathogen, intestinal inflammation ensues. When *Salmonella* is alone in the gut with no other types of microbes around, it is forced to rely on the fermentation capacity encoded within its own genome. But *Salmonella*'s genome contains few of the scissor-like enzymes required to chop up dietary MACs and mucus that many beneficial microbes have. As beneficial gut microbes degrade and consume their meal, they create a buffet of waste products. In a healthy, colonization-resistant microbiota, competition for resources is intense and by-products from one microbe are rapidly consumed by another, leaving nothing behind for invaders like *Salmonella* to steal.

Resource stealing works best when the natural consumers of these resources in the gut have been crippled, and antibiotics do precisely that. Antibiotics disrupt connections within the complex food web, leaving a resource gap that *Salmonella* and *C. difficile* can exploit. But the pathogenic party crashers need to be present—lying in wait, as it were—to take advantage of the newly hospitable landscape. So if you've recently been prescribed antibiotics, it would be a good idea to avoid situations that are associated with *Salmonella* exposure, such as ordering eggs at a restaurant or playing in a public sandbox. Over time, the gut microbiota can recover, more or less, from the antibiotic treatment, reestablish an efficient food web, and resist pathogen invasion.

The success of many pathogens requires a sophisticated strategy for invasion of the gut ecosystem, but also a plan for persistence. What's the point of crashing a party if you don't have a chance to stay around long enough to enjoy the spoils? *Salmonella* achieves persistence by disrupting the gut environment in a way that ensures its continued access to resources. By disruption, we mean causing diarrhea and intestinal inflammation. This radical change in the gut environment makes it more difficult for friendly microbes to regain a foothold and recapture the resources that *Salmonella* is taking. In changing the ground rules of how the gut ecosystem operates, these bad characters gain an ongoing advantage over the good guys.

We tend to think of immune system responses to a pathogenic attack such as inflammation as helping the body eradicate unwanted invaders. But some pathogens have figured out how to trigger an immune response that actually works in their favor. *Salmonella* has to face the initial hurdle of expanding within the microbiota, something that antibiotic-induced microbiota damage can help it achieve. But once it has effectively cleared this hurdle, its next goal is to trigger inflammation in the intestine. The resulting inflammation changes the rules of life within the gut, to *Salmonella*'s advantage. *Salmonella*, like many pathogens, is a master at shaping and subverting the host's immune response and physiology to its own advantage.

Strength in numbers is just one protection that the microbiota provides. As our understanding of how the microbiota helps to regulate the immune system becomes more sophisticated, it is becoming clear that colonization resistance is not just a game of exclusion. The microbiota is engaged in an ongoing dialogue with the gut. Microbes can coax the immune system to launch an appropriate response large enough to deal with the threat but not so big that it sets off an autoimmune reaction or causes excessive damage. Some microbes in the gut take on a more direct role in dealing with pathogens by secreting their own set of pathogen-targeted antibiotics. As opposed to the carpet bombing approach that high doses of oral antibiotics provide, the antibiotics deployed by residents of the microbiota appear to cause much less collateral damage.

Besides decreased colonization resistance, there is a second real danger that accompanies too much antibiotic use: the so-called antibiotic-resistant superbug. The indiscriminate use of antibiotics has helped to create highly armored pathogens that can survive attack by many of the world's most powerful antibiotics. How were these microscopic Frankensteins created? When a group of bacteria, like those in the gut, are exposed to antibiotics, one will occasionally by chance harbor a genetic trait that renders the microbe resistant. That one bacterium, because it survives, can proliferate even in the face of the antibiotic and give rise to an army of mutant resistant bacterial species. Since bacteria are very adept at sharing genes in a process known as lateral gene transfer, an antibiotic-susceptible bacterial species in close proximity to the antibiotic-resistant microbe can absorb a copy of the highly coveted resistance gene and gain an advantageous characteristic: antibiotic resistance. You could imagine a scenario in which after multiple courses of antibiotics, resistance genes become more and more abundant in the gut. If a pathogen happens to be transiting through the gut and picks up one or more resistance genes from the microbiota, a potential superbug is born.

Infection with a multi-drug-resistant microbe is a nightmare situation without a good solution. For the first time since the advent of antibiotics, people are dying from bacterial infections that would have been curable if it wasn't for these pathogens' extreme resistance to available drugs. New antibiotics can be developed to address this issue, but they won't help us escape the arms race we have entered with pathogenic bacteria. To stay one step ahead of these pathogens, we need a multipronged approach. First, we need a constant supply of novel types of antibiotics, ones that bacteria have never experienced and therefore have not yet developed resistance to. Second, we need to strengthen our own internal defenses by building and maintaining a strong and diverse microbiota, which will help to minimize the use of antibiotics in the first place.

GOING WITH THE FLOW

Dr. Purna Kashyap of the Mayo Clinic sees many patients with gastrointestinal motility issues, such as recurrent diarrhea or constipation. There are a number of illness in which these gut motility problems manifest, such as inflammatory bowel disease (IBD) and irritable bowel syndrome (IBS). Dr. Kashyap was concerned that these chronic issues would disrupt the microbiota and exacerbate the underlying condition. When he started working in our lab in 2010 there was a paucity of information about how microbes in the gut are affected by changes in intestinal transit.

The gut, as noted earlier, is like an internal bioreactor. It is filled with contents like food and water that flow through the tube and are processed by the human cells and the microbiota. The rate at which these contents flow through the gut can drastically change the environment the microbiota experiences. If flow rate is very rapid, the microbes have limited time to consume food passing through and may be more easily washed out of the gut. Alternatively, a gut in which the contents transit very slowly would create a different set of challenges for the microbiota. In both cases of extreme deviation from normal gut transit time, there is a risk that the health of the microbiota could suffer.

Dr. Kashyap wondered if people with persistent diarrhea, where flow is too fast, or constipation, where flow is too slow, could experience a failure in their internal bioreactor. He found that both diarrhea and constipation change the environmental conditions within the gut. Microbes that are better adapted to fast-moving intestinal contents become more abundant when the rate of flow is fast. Conversely, those that do better when motility is slow thrive in a constipated gut. In both situations, when flow rate is too rapid or too slow, the diversity of the microbiota drops. This situation has a destabilizing effect, leaving resources available for an invading pathogen. So antibiotics are not the only way to disturb the microbiota and decrease colonization resistance.

Diarrhea, antibiotics, and potentially other disturbances can set off a vicious cycle. Infection by one pathogen often increases gut motility and further disrupts your microbiota, making you more susceptible to acquiring another intestinal pathogen. In the case of *C. difficile*, FMTs offer a crude but effective escape. But if you had to get a fecal transplant, who would you pick as your donor? And what are the important considerations for effectiveness and safety?

DON'T TRY THIS AT HOME

In 2013 the FDA announced that it would regulate FMTs in much the same way that an experimental drug is regulated. After patients and the medical community voiced serious concerns that the FDA's ruling would effectively restrict access of a lifesaving treatment, the FDA revised its stance to open access to FMTs for treatment of those with recurrent *C. difficile* disease.

If this situation sounds like another example of too much regulatory red tape, there is yet more to this story. At the same time that the FDA is restricting FMTs for everything other than *C. difficile* infection, it is not requiring that fecal material used for transplants be subjected to a standardized set of safety tests to make sure that it is free from infectious agents. Most physicians do conduct safety screening of FMT donor samples; however, the nature and extent of testing is dependent on which medical facility is performing the FMT. Presently there is no consensus for which tests are necessary for FMT donor stool. Obviously, stool material needs to be clear of infectious pathogens like HIV, parasites, and other types of microbes that could transfer disease. But is that enough? Because the microbiota can confer physical and psychological traits in lab animals and perhaps also in humans, should all donors be thin, free of psychological disorders, and not suffering from allergies? How important is it that the donor was born vaginally, was breast-fed as an infant, has not had multiple rounds of antibiotics, and eats a plant-based diet? Finding a microbiota donor who could satisfy all these qualifications could be difficult.

Newly formed companies are already providing prescreened fecal material to hospitals for FMTs. These companies' operations are similar to those of a blood bank. They collect fecal material from donors who have had safety tests and they generate revenue through fees they collect from hospitals that use their material for FMTs. This saves the hospital from having to identify and screen donors individually. However, now the FDA has cited safety concerns about these stool banks, and new regulations are likely to result.

FMTs are a simple solution to a highly complex problem. An FMT provides an effective reset button, allowing a diseased microbiota to regain a healthy state. Why shouldn't this approach translate to other diseases in which the microbiota is disrupted, such as IBD, autism, autoimmune diseases, even obesity? Imagine that the solution to the obesity epidemic was as easy as a fecal transplant. Unfortunately, FMTs are not turning out to be the cure-all people had hoped for.

A small-scale clinical trial investigating the effectiveness of FMTs to improve obesity-related comorbidities has shown some promising, although not stunning, results. Obese people receiving an FMT from a lean donor had a temporary improvement in their insulin resistance, but did not lower their body mass index or percentage of body fat. Treating inflammatory bowel disease with FMTs also has not resulted in the high cure rates observed for CDAD. And there were more substantial side effects, such as fever and painful bloating, in some patients who had no remission in their disease after FMT. All these studies are still preliminary, so the jury is still out on whether FMTs could improve outcomes for some people with IBD or IBS. Several large-scale clinical trials are currently under way to determine what kind of future FMTs will have outside of treating CDAD.

Microbiota-linked diseases are complex and are distinct from one another. Some diseased microbiotas, like those accompanying *C. difficile* infection, have very low biodiversity and resemble a fairly barren landscape. In such a case, sowing new seeds has a good chance of successfully changing the overall habitat. Alternatively, if the ecosystem is similar to a yard full of established but unwanted weeds, it may be difficult to correct the problem by dispersing desirable seeds onto an overgrown landscape. Either thorough weeding prior to reseeding or eliminating the conditions that allow the weeds to thrive would improve the chances of restoring the desired plants. In this case, clearing the gut of unwanted microbes via antibiotics or an enema could maximize the effectiveness and durability of the newly transplanted microbiota. Special fertilizer, in the form of dietary MACs, could be applied to encourage the growth of the new beneficial organisms. These two possible scenarios for the diseased microbiota—the barren landscape or the weed-filled yard—are simplifications of the vast number of possible states a malfunctioning microbiota can take. As the details of each disease are better understood, tailored strategies to enable FMT success will become apparent.

ENDING THE DARK AGES OF FECAL TRANSPLANTS

What if it were possible to transfer curative beneficial microbes into an unhealthy microbiota without the possibility of introducing infectious agents or increasing susceptibility to other diseases? Surely the crude form in which FMTs are performed today is just the beginning of what could be a revolution in treating diseases in which the microbiota has been attacked or is in disrepair. What does the future of curing the sick microbiota look like?

One way to minimize the risk associated with FMTs would be to have your own stool stored should you need a transplant in the future, the way a patient's blood can be stored before surgery. An FMT with your own stool would remove the worry of passing new infectious agents from one person to another. North York General Hospital in Toronto, Canada, has started a pilot program to collect stool from patients as they are admitted. Like backing up your computer's hard drive in case of a crash, in most cases the reserve would not be needed, but could be critical if the patient develops CDAD as a result of his or her stay in the hospital. Because of the high use of antibiotics and prevalence of drug-resistant strains of *C. difficile* in the hospital environment, hospitals can be breeding grounds for CDAD. By banking a fecal sample for each patient, physicians would have access to FMT material without the cost of donor screening and the possibility of inadvertently spreading disease. Preemptive stool banking for patients definitely seems like a prudent course of action at this point and may become routine in more hospitals.

Even with stool banking, there are still major issues around FMT. While it does not expose the recipient to new infectious agents, fecal material still needs to be handled by health-care workers in preparation for the transplant, exposing the technician to any infectious agent that might be present in the sample. And all joking aside, smell is a significant issue in dealing with fecal material. One way to skirt these issues would be to use a defined mixture of microbes that could be grown in a laboratory instead of using fecal material. Use of this manufactured product would ensure that no infectious microbes were present and would standardize treatment.

A group of scientists has created a mixture of thirty-three bacteria they named RePOOPulate to treat CDAD. The researchers selected these specific bacterial types because they are known to be beneficial and are often depleted in the microbiota of individuals with CDAD. The bacteria were even screened for antibiotic resistance to ensure they wouldn't transfer any resistance genes to other microbes in the gut. Small tests of RePOOPulate transplants on CDAD patients have shown promising results. Two patients were cured by the microbial cocktail and even six months later both microbiotas still harbored the thirty-three RePOOPulate bacterial strains. In fact, those bacteria made up a quarter of the total strains found in either microbiota. Not only had these bacteria successfully colonized the gut, they were persisting. Amazingly, both patients later received multiple courses of antibiotics for infections unrelated to their CDAD and in both cases the CDAD did not return. The new RePOOPulate microbiota was either keeping *C. difficile* in check when the antibiotic disruption occurred or had completely eradicated it.

The preliminary success of using defined mixtures of microbes instead of crude stool samples points to a future of more targeted approaches. There are a number of fledgling companies that are taking the bet by designing custom mixtures of microbes to be used for microbiota transplants. Some of these mixtures are even being packaged in pill form, which would also drastically reduce the cost and risks associated with delivery by enema or nasogastric tube. These "crapsules," as they are sometimes called, could become part of a routine postantibiotic regimen to mitigate microbiota damage.

Although bacteria are cleaner than fecal material, treating people with living organisms still makes regulators uneasy. Ingesting live organisms could be like opening a Pandora's box within the gut. The fact that the microbes from the RePOOPulate study were still found in the gut six months after treatment may mean that if something goes wrong, getting rid of these microbes might not be so easy. Drug companies work under the paradigm that molecules, not living things, make the best medicine. Molecules can be easily regulated, patented, and, because they are not alive, dosage is easier to control.

In the RePOOPulate study, what would have happened if these bacteria comprised half of the microbiota or more? Could that cause a problem and, if so, how could their numbers be reduced? These living organisms can multiply on their own after being administered, which is likely a requirement for an effective treatment. But our lack of control over how much they multiply makes proper dosing difficult compared to a conventional chemical drug, in which dosage can be dictated with precision.

Microbes in the gut are synthesizing molecules (remember our unsupervised drug factory) that influence inflammation and repair defects to the intestinal lining, among many other effects. Another approach that avoids having to administer living organisms involves using microbiota-produced chemicals. Such an approach would be similar to repairing the soil in the hopes that the flower seeds that are already present could once again grow and flourish. Many new microbiota-derived molecules are likely to be discovered in the coming years.

UPDATING YOUR INTESTINAL OPERATING SYSTEM

The initial triumphs of fecal transplants and microbiota-based therapies have invigorated the field of microbiota research as scientists and clinicians come to grips with the power that poop, or more precisely its inhabitants, can have to cure previously intractable diseases. Over the coming few years, strategies to reprogram the gut microbiota will extend well beyond the introduction of stool or alterations in diet. Pharmaceutical companies are starting to look for drugs that can target members of the microbiota to alter community composition or function. Engineered microbes that can detect disease or deliver a drug will also become part of the arsenal used for microbiota reprogramming.

While tools continue to be developed, the current focus is on

FMTs and much remains to be learned. It is not yet clear how easy it will be to replace a bad microbiota with a good one. Although a *C. difficile*–infested microbiota is easily rebooted with healthy stool, the replacement of an obese person's microbiota with that of a lean donor is not so straightforward. Early reports indicate that the lean microbiota only takes over for a short period of time, and that within three months the community reverts back to the original, obese microbiota. Failure of the obese recipients to change their diet may have been an important factor in the inability of the donor microbiota to persist. As noted earlier, studies in mice convincingly show that reinforcement of a lean microbiota with a diet high in fruits and veggies is essential to overrun an obesity-promoting microbiota. A diet that helps nourish these lean microbes allows them to effectively invade the obese microbiota and protect the mice from weight gain, changing the hosts' destiny. Perhaps in the human study, if the recipient of the lean microbiota had adopted the dietary habits to support this community, the new transplanted microbiota could have persisted and improved health. While we know that improvement in diet can help obese individuals lose weight and decrease their risk of obesity-associated conditions, perhaps dietary intervention alongside FMTs can provide a jump start to a failing ecosystem. This one-two punch of a new microbiota plus dietary reinforcement could be an answer to this seemingly intractable problem.

Even without introducing a new set of microbes, diet can aid in eradicating intestinal pathogens. Shigellosis, an infection found primarily in the developing world, is caused by the bacterium *Shigella* and is typified by bloody diarrhea. Antibiotics are the usual course of action to control the infection, but researchers have found that if cooked green bananas are given in addition to antibiotics, patients regain their health much faster. In this case, the cooked green bananas, and specifically their MACs, act as a fertilizer to the decimated landscape, promoting the growth of good species. Similar to FMTs, this treatment helps to restore a healthy microbiota to eradicate the pathogenic bacteria, but uses a dietary strategy to do so. It also serves as an important reminder that diet is a powerful and accessible tool that enables everyone to take some control in programming (or re-programming) the microbes that govern so much of our biology.

CHAPTER 8

The Aging Microbiota

OUR LIFELONG COMPANIONS

The fight against aging drives a huge industry. We submit to torturous procedures like Botox injections, acid peels, and microdermabrasion in the hopes of regaining a more youthful appearance. We perform mental gymnastics, completing countless sudoku puzzles and online brain-training games, trying to keep our minds sharp. Working to increase flexibility through yoga and to maintain muscle mass by doing weight-bearing exercises, we try to slow the physical decline of our bodies and our health. However, new studies reveal that there is another critical component to maintaining youthful vigor: nourishing your aging microbiota. Like all other physical and mental aspects of the human body, the microbiota also shows age-related wear and tear over time. How rapidly it declines can predict how rapidly your health will follow suit. But just as there are ways to combat (or at least delay) the aging of our skin, mind, and body, there are ways to keep your microbiota more youthful.

The collection of bacteria residing in the gut is amazingly constant over time. Of course there are short-term variations with some bacterial species blooming and then retreating from one day to the next. Often these fluctuations can be explained by external factors like antibiotic use, diet change, even fever. But sometimes the causes of these deviations are more cryptic. Regardless of the minor disturbances that occur, if you sampled your microbiota now, and then five years from now, the composition would still be recognizable as uniquely yours. Each person houses a core set of species that is stable over time, similar to your other physical traits, like eye or hair color. The people who share a microbiota composition most like yours are your closest relatives, forming an internal family resemblance of sorts. The core bacterial strains that are part of your microbiota fingerprint make up anywhere from one third to two thirds of the total species in the gut and remain with each of us for decades. Scientists believe that many of these core bacterial species could be our companions for a lifetime, much in the way that we are stuck with the nose we were born with (plastic surgery notwithstanding). And like our nose, there is evidence that some of these species are acquired directly from our parents and are common to our siblings. Some may be acquired at birth or early childhood and persist for our entire lives—microbial traits inherited from generations past. Added to these stably associated core microbiota inhabitants are species that are more fluid, varying over time, much in the way that our hairstyle or clothing choices change with time. But these changes don't mask the unmistakable makeup of an individual's microbiota over the long term. The shared family resemblance in microbiota is much weaker, by comparison, and strangers typically share even fewer bacterial "traits."

With all the exposure to different environmental microbes, diet fads, and antibiotics that occurs throughout our lifetime, it is hard to imagine that a portion of our microbiota is so durable. It seems as if some species of bacteria, once they move in, can keep other, similar species from displacing them. Each species of bacteria has a range of niches or "professions" that it carries out in the gut. Some bacterial species have a broad skill set, meaning that they can occupy diverse niches within the gut. A bacterium that consumes pectin will thrive after an apple is eaten. But if that same bacterium can also subsist on carbohydrates found on the intestinal wall, it can survive there or maybe even proliferate, even when fruit is not available. This bacterium has a diverse niche and can adjust based on diet or on which bacterial competitors may be present. Some bacterial species, however, carry a more specialized toolbox. If a bacterium really specializes in pectin consumption it will become more abundant after the apple is eaten, leaving little room for any other pectin-degrading bacteria that might have been swallowed along with the apple. In this scenario, the species that has already set up camp in the gut wins.

Even in the face of huge challenges like pathogenic infection or antibiotic treatment, some gut bacteria can hide in small "caves" called crypts located along the side of the intestinal wall. Once the threat passes, these hunkered-down microbes serve as a potential reservoir to reseed the gut. These types of strategies, completely filling niches within the gut to ward off bacteria with similar metabolic strategies and laying low while outside threats ravage the microbiota, result in a resilient ecosystem capable of delivering stability over time.

But aging is not an acute stressor like a pathogen or antibiotics. The long-term system-wide deterioration that typifies aging also occurs within the microbiota—a deterioration that can impact our overall health.

THE RETIREMENT COMMUNITY

For the microbiota, the aging gut can be a place of dramatic environmental shifts. The speed at which food transits through the digestive tract lessens, which can lead to chronic constipation. Age-related decline in our sense of smell and taste and a decrease in our ability to chew can dramatically change our diet to one deficient in fibrous plants and chewy meat. More time spent in hospitals and on antibiotics can increase the chance that a pathogen like *C. difficile* rears its ugly head. All these factors produce an environment in the gut that might be drastically different from what it was in our younger years. Increased flatulence, a common complaint among the elderly, is one giant clue that the microbiota is undergoing a substantial reorganization.

The scientific understanding of the microbiota changes that occur near the end of life is, ironically, immature. However, recent studies have begun to unravel the mystery of what happens to the microbiota as a person ages. In 2007, a group of researchers at University College Cork in Ireland started ELDERMET, a project designed to look at the relationship between diet, microbiota, and the health status of several hundred people over the age of sixty-five. Their results offer clues about how the microbiota ages, what age-related microbiota decline means for our longevity and health, and how we can reinvigorate the microbiota during this critical period in our lives.

The ELDERMET study found that the microbiotas of seniors are very different from one another, contrasting with the relative similarity of young adults' microbiotas. This situation mirrors what happens at the beginning of life, when the forming microbiota of babies is more chaotic. You can think of each person's microbiota throughout life as a grain of sand in an hourglass. At the top of the hourglass, like the beginning of life, the sand occupies a large area. The big distances separating some of the grains represent the big differences that are observed when comparing the microbiota of two infants. By five years of age and throughout adulthood the composition of each person's microbiota converges, much in the way that the neck of an hourglass pushes the grains together. As we get older, each person's microbiota diverges from others', spreading out like sand at the bottom of the hourglass.

When scientists looked more carefully they found that the differences in microbiota composition between study participants were not randomly distributed. They found three different clusters of microbiota composition. One cluster was from people still living in the community, whose microbiota resembled that of younger people from the same area. The other two clusters were from people residing in day hospitals and in long-term-care facilities. It seems that, when it comes to microbiota composition, location matters. But what is it about the location that is important? When people make the transition from independent living to a long-term-care facility, their diet also undergoes a transition. The higher-fiber diet that typified the community dwellers and those in the day hospital gives way to a diet that is markedly lower in fiber in the long-term-care facility. While the reason for the fiber-poor diets in these facilities is not clear, decreased fiber is common in food prepared in cafeteria-like settings, and the problem may be compounded by a desire to make food easy to chew for the elderly. This difference in fiber consumption mirrored the differences seen in microbiota composition. People on a low-fiber diet had a lower-diversity microbiota, whereas a high-fiber diet made for higher microbiota diversity. The higher-fiber-eating community elders also had more health-promoting SCFAs, lower markers of inflammation, and in general were in better health.

These microbiota differences could be a result of the fact that people with poorer health are more likely to be living in long-term-care facilities in the first place. What is the relationship between diet, microbiota, and health? Does microbiota decline result from, or lead to, health decline? Could it be that less-healthy people end up in long-term-care facilities, eat a low-fiber diet, and microbiota deterioration results? This is another chicken-and-egg issue requiring more studies to resolve. But based on what we know about the importance of the microbiota to health, even if the microbiota change is secondary to health decline, its apparent deterioration is likely to at least compound health problems.

The ELDERMET scientists have a hunch that diet decline sets off this whole cascade. The key clues came when they looked at the order of diet change, microbiota change, and health deterioration. When first entering a long-term-care facility, people had a diet that was different from that of others who had lived there for more than a year. But after a month the new residents' diet changed, becoming more similar to the longer-term residents'. The microbiota of the new arrivals, however, took up to a year to resemble the microbiota from residents who had been in a care facility for an extended period of time. While dietary change can lead to rapid alterations in the microbiota, the core lifelong microbial residents can slowly be lost when microbiota accessible carbohydrates are diminished. In this case, diet change happened first and the microbiota change followed. Most important, measures of frailty corresponded to the biggest differences in the microbiota shift. The data from the ELDERMET study points to a chain of events in the elderly that starts with diet deterioration followed by microbiota change and then health decline.

You might imagine that pharmaceutical companies would be interested in making a pill that consisted of "young" microbes that could be added to an old, deteriorating microbiota. Unfortunately, when dealing with a complex community like the microbiota, answers to problems tend to be equally complex. Similar studies performed in elderly populations from Italy, France, Germany, and Sweden found microbiota differences between old and young, as in the Irish study, but they were not the same differences. The "old" bacteria found in the Irish were not the same ones that bloomed in elderly people from other European countries. Similarly, there did not appear to be many consistent "young" bacteria between regions.

Clearly a simple find-and-replace strategy is not so straightforward for keeping a youthful microbiota. But in retrospect these differences should not be that surprising. Each geographically and culturally distinct population would be expected to house a somewhat unique microbiota, one that is a product of eating a specific diet and encountering certain types of environmental microbes. How each microbial community ages is likely to take its own trajectory. A valuable implication of these studies is that diet is an important factor in maintaining your youthful microbes as you age.

INFLAMMAGING

We know there is a link between the state of the microbiota and frailty during aging. Specifically, a more diverse microbiota is associated with better health parameters like lower inflammation, greater muscle mass, and less cognitive decline. But it is not clear how this connection works. What is it specifically about the microbiota that leads to better aging? If we could understand the *how*, we would be on the road to using microbes to improve age-related health decline.

As the human body ages, almost every aspect of its biology declines: the kidneys don't filter out toxins as well as they once did, the heart is weaker from years of mechanical stress, even the brain's ability to recall memories fogs over time. But one of the more striking degenerations involves the immune system. The immune system takes a beating over our lifetime, hits that range from the nonstop vigilance it musters to scout out potential pathogens to the damage it sustains with each skirmish it engages in. Some of this damage is never fully repaired and over time accumulates to weaken the overall functioning of the immune system. This age-related decline is referred to in scientific circles as immunosenescence. It happens to all of us and even to our pets; no one is immune, so to speak.

Immunosenescence is highly complex and involves all branches of the immune system, but one of its manifestations is a low-grade chronic inflammation referred to as inflammaging. In inflammaging the balance between pro- and anti-inflammatory responses within the immune system is tilted toward the pro-inflammatory side. This inflammatory state has been linked to many age-related diseases such as dementia, Alzheimer's disease, and arthritis and can negatively influence the microbiota. Since many of the microbes that enjoy a slightly inflamed gut are also capable of perpetuating inflammation, this age-related pro-inflammatory slant could set off a self-sustaining cycle of microbiota decline that feeds inflammaging and deteriorating health. Sprinkle in decreased fiber consumption and lessen exercise over time and the potential for an aging microbiota to exacerbate health decline is great.

One universal feature of the aging microbiota is the increased abundance of pathobionts, bacteria that are capable of behaving in a pathogenic or disease-causing way. We all have pathobionts lurking in our microbiota. Under normal, healthy circumstances these bacteria behave in a benign manner, or are not abundant enough to cause problems. However, when the gut is inflamed, pathobionts can expand and help perpetuate inflammation. Pathobionts, while a common feature of the older microbiota, can also thrive under certain dietary conditions. Laboratory mice fed a diet rich in saturated animal fat show an expansion in the amount of pathobionts in their microbiota. This increase in pathobionts was not observed in animals fed an equivalent amount of plant-based polyunsaturated fats.

How can one minimize inflammaging and escape this cycle? By consuming a diet rich in MACs and restrained in saturated fats from animal sources. Increased dietary fiber and lower fat consumption among the elderly is correlated with greater SCFA production and less inflammation in the gut. As microbes ferment MACs from dietary fiber, they produce SCFAs that can decrease inflammation.

Similarly, a low-fat diet can minimize inflammation by discouraging the proliferation of pathobionts. A healthy gut environment is detrimental to pathobionts, which perpetuate inflammation and rely upon it to thrive.

FITNESS FOR YOUR MICROBIOTA

In the West we typically commute by car to our sedentary jobs and come home at the end of the day too exhausted for much beyond watching TV on the couch. Hitting the gym a couple of times a week on the way home from work, while beneficial, is not enough to compensate for the overall lack of activity in our modern lifestyle. As we age, getting adequate exercise can become more and more challenging. Our bodies don't adapt as easily, we are less flexible, and we're less motivated. But an overwhelming amount of evidence points to the importance of maintaining physical fitness as we age. Exercise can delay the aging process and reduce the risk for many debilitating diseases including obesity, heart disease, cancer, diabetes, and even depression. Activity forms a counterbalance to eating by burning calories, strengthening the heart, improving mood, and ameliorating age-related physical decline. But it also appears that exercise can help minimize the effects of immunosenescence and inflammaging, and even might affect the microbiota.

As people age they become more sedentary, which contributes to slower gut motility. We already know that intestinal transit time impacts the gut environment and the composition of the microbiota, so it makes sense that quickening transit time through physical activity could alter the microbiota. But to carefully address how exercise is tied to the microbiota is complicated because individuals who exercise often consume a healthier diet, making it difficult to disentangle the effect of these aspects of lifestyle on the microbiota. Studies in laboratory mice, which allow for more independent control of diet and exercise, show that each factor on its own can improve the state of the microbiota. But combining a better diet along with exercise has the greatest potential to impact the microbiota and our health.

OUR MICROBIAL ALLIES IN THE "WAR ON CANCER"

Cancer, typified by uncontrolled cell growth, is, in many ways, a disease of the immune system. Cancer cells arise spontaneously in our bodies throughout our lives but normally our immune system finds them and stamps them out. In cases where this seek-and-destroy strategy fails, the cancer can grow and spread. The longer these malignancies are allowed to proliferate, the "smarter" they can become, exploring many different strategies to mask themselves from the immune system and multiply with abandon, until they find one that works. Tumors can create safe havens by crafting microenvironments that exclude the immune cells patrolling the body for malignant growth. These protected microenvironments allow the cancer cells to grow safely shielded from the immune system.

Some cancer treatments rev up the immune system to better root out the camouflaged cancer cells. One such treatment is a chemotherapeutic called cyclophosphamide. This drug invigorates the immune system and blocks the tumor's ability to recruit blood vessels, which are critical supply lines of nutrients. One of the unforeseen effects of this drug is its ability to make the gut lining slightly porous, allowing bacteria from the microbiota to escape the confines of the intestine. Scientists studying mice that were given cyclophosphamide found gut microbes in the spleen and lymph nodes of these mice. At first glance this may seem like a disastrous outcome of the chemotherapy—loose microbes roaming the body and invading other tissues. However, instead of causing problems, these microbes actually helped. By being in tissues where they don't belong they alerted the immune system and the immune system responded to the alert by launching an attack. And due to the inevitable lack of precision of the immune system's attack, some of the response was targeted toward the cancer cells and resulted in the tumors shrinking. If the mice were pretreated with antibiotics before cyclophosphamide, the treatment's success was blunted. Antibiotics crippled a much-needed accomplice, the gut microbes, making immune system activation and the anticancer response less robust.

How does this study relate to the treatment of human cancers? Many therapies, while aimed at destroying cancer cells, often cause a lot of collateral damage. Much of this damage is sustained by the immune system, rendering patients highly susceptible to opportunistic infections. So to decrease this possibility, antibiotics often accompany cancer treatment prophylactically. But the positive role of immune system stimulation caused by our resident microbes may have clinicians rethinking this practice. When drugs work through augmentation of the immune system, the state of the microbiota needs to be considered. Differences in the microbiota could also at least partially explain disparity in how people respond to immunotherapy for cancer and other diseases.

Clearly, when targeting the immune system with therapies, we need to be cognizant of the microbiota. But not all cancer therapies work through immune system activation. Radiation and certain chemotherapies work by preferentially killing cells that divide more rapidly, a characteristic of cancer cells. Unexpectedly, these types of therapies can also be impacted by the microbiota.

A group of researchers investigated two chemotherapy drugs called cisplatin and oxaliplatin. These drugs are used to treat a variety of cancers, including colorectal cancer, lymphomas, and sarcomas. They work by throwing a metal wrench (these drugs even have a platinum core) into the cell's replication machinery, effectively stopping cells from dividing. Because cancer cells replicate faster than normal cells, these drugs can put the brakes on the runaway growth of malignant cells (as well as some other, faster-growing normal cells like those that make hair). But a second phase required for these drugs to work involves the immune system sweeping away the stalled cells. To do so, the immune system needs to be able to access these cells, which can lie hidden deep within the microenvironment created by the tumor.

The tumor microenvironment can be quite fortified, but it is not Fort Knox. An immune system that is tuned to an aggressive set point can penetrate this environment. If there is one thing the microbiota excels at, it's tuning the immune system. Scientists wondered if the microbiota could spur the immune system to eradicate cancer cells in the tumor microenvironment. They found that mice on antibiotics had a tumor microenvironment that looked more hospitable to the malignant cells and less likely to be infiltrated by the immune system. When the scientists gave the tumor-bearing mice the platinum-based chemotherapy, the drug didn't work as well in the mice pretreated with antibiotics. In this case, unlike with cyclophosphamide treatment, microbes weren't transiting out of the gut into other tissues, but instead were tuning the immune response to the cancer from within the gut. When a complete, healthy microbiota was present, the immune system could efficiently infiltrate the tumor microenvironment and clear the jammed-up cancer cells.

It is important to keep in mind that because these studies were performed in mice, additional work is needed to ascertain how these results would translate to human cancers. However, influence of the human microbiota on the effectiveness of cancer drugs merits further investigation. The ecosystem composed of our human cells and our collection of microbial cells is incredibly intertwined. We should expect that perturbing one part of the system through drugs could produce unforeseen consequences.

Antibiotics given alongside chemotherapy to mitigate the risk of infection may not be, in some cases, the most prudent course of action. Perhaps in the future, instead of prescribing microbiota-damaging antibiotics, therapy-enhancing bacteria might be given to patients to improve chemotherapy outcomes. As the understanding of individual variations in microbiota composition increases, chemotherapy could be tailored not only to the type of cancer but also to a person's microbiota "type."

Similarly, it is possible that the microbiota, through its connection with the immune system, may affect a propensity for cancer, or its progression. Are there microbes that are cancer promoting? Can improving the health of the microbiota reduce our chances of getting cancer or lessen the aggressiveness of an existing cancer? We don't know the answers to these questions yet. But when thinking about ways to prevent or rid our body of diseases like cancer, we need to take an approach that minimizes damage to the beneficial aspects of our biology, like our microbiota. For example, when confronted with an ant problem, instead of using a broad-spectrum pesticide to kill every living insect within a ten-foot perimeter of your home, limiting chemical use while nurturing natural ant predators like spiders, wasps, and beetles can keep an infestation at bay. Similarly, for diseases like cancer, coupling treatments such as chemotherapy and radiation with measures to improve the ability of the microbiota to boost the immune system may provide a more durable response.

THIS IS YOUR MICROBIOTA ON DRUGS

The microbiota exerts influence over the action of a number of pharmaceuticals. And as the microbiota is studied in relation to more and more medicines, this list will grow. The microbiota can directly affect the strength of some drugs, which can make these drugs more or less effective in certain individuals, depending on the particulars of their microbiota. Other therapeutics are affected by the microbiota in a more indirect fashion. The uniqueness of each person's microbiota and how it interacts with certain drugs can be a source of variability in drug effectiveness and side effects.

As we age we encounter more and more health issues that require the use of medications. This reality makes it important to understand that by taking medication, we are introducing a variable into a highly complex system that consists of numerous still uncharacterized interactions between human cells and microbial cells.

Acetaminophen, more commonly known as Tylenol, has been a widely used pain reliever and fever reducer since the 1950s. Many of the details of how this drug operates on a molecular level within the human body are known, but less understood is why it can produce huge variations in response and side effects among people. Acetaminophen overdose is the leading cause of acute liver failure in the United States, but the basis for failure is unexplained in about 20 percent of the cases. Filling in this missing piece of the puzzle is critical to ensure that each person receives a dose that is both effective and safe. One variable that can affect drug dosage is the speed at which it is eliminated from the body. If a drug is cleared quickly, then a higher dose may be needed to maintain adequate amounts of the drug in the target organ or circulating in the blood. However, if the drug lingers around the body longer than expected, the risk of adverse side effects and even overdose increases.

Part of what decides the speed at which a drug is excreted from the body is how rapidly it is processed by the liver. The liver serves as our body's chemical detoxification system. Chemicals in our bodies that arise from things we ingest, the medications we take, our own cells' metabolism, or the metabolism of our microbiota are handled in a conveyor-belt-like fashion. Molecular machinery within the liver attaches chemical "tags" to potentially harmful compounds to facilitate their passage out of the body. A person's genetic makeup and the quantity of other chemicals circulating throughout his or her system can impact how rapidly chemicals are processed. If there are many chemicals awaiting tags, a "backlog" can develop. But instead of the chemicals waiting in a queue-like fashion in the liver, they continue to circulate throughout the bloodstream until there is adequate capacity in the liver to modify them. The additional time spent in circulation translates into increased exposure to the effects of these chemicals.

Drug manufacturers and physicians take excretion time into account when creating drug formulations and determining what dose to prescribe. If a drug typically is modified and excreted rapidly, a higher dose is required for that drug to have the desired effect. However, individual variations in how drugs are tagged and excreted can result in situations where a person receives a higher or lower "effective" dose of a drug than intended. Say you are prescribed a drug to treat a certain condition. If you happen to be a person who tags and excretes the drug more slowly than the typical rate, you would experience a higher dose of the drug, which could cause a greater risk of adverse side effects. If on the other hand you excrete the drug too rapidly, your condition may not be treated effectively. Outside factors can also impact the speed at which drugs are processed in our bodies. One familiar example is that individuals taking certain medications, such as statins, are cautioned against eating grapefruit. Naturally occurring chemicals in grapefruit can directly compete with statin processing in the liver, slowing down statin excretion and inadvertently raising the dose to potentially harmful levels. But food-derived chemicals are not the only source of variability. Chemicals produced by the gut microbiota can also affect how our bodies metabolize drugs.

When studying how quickly acetaminophen is eliminated from the body, scientists found that a waste product generated by the microbiota, *p*-cresol, matters. The amount of *p*-cresol produced by a given microbiota depends on the types of microbes found in the gut and the amount of amino acids (the building blocks of protein) consumed. Microbes within the gut metabolize amino acids and form *p*-cresol as a waste product. Once formed in the colon, *p*-cresol is absorbed into the circulation and needs to be tagged by the liver and excreted. Since the liver enzyme responsible for tagging *p*-cresol is the same one responsible for detoxifying acetaminophen, an overabundance of *p*-cresol could result in a backlog in acetaminophen processing. A person with a microbiota that produces a large amount of *p*-cresol could effectively experience a larger dose of acetaminophen than someone whose microbiota is a low *p*-cresol producer. Even if you could determine whether your particular microbiota had the capacity to produce a lot of *p*-cresol, how much was produced on a given day may also depend on your diet, specifically how much protein you recently ate.

How the microbiota impacts our body's reaction to acetaminophen is somewhat indirect, affecting the rate at which this drug is excreted. But some medications are directly affected by microbes in the gut. Digoxin, a drug used to treat cardiac irregularities, is a variant of digitoxin synthesized by the foxglove plant, which has been used as a medication for heart conditions for hundreds of years. Digitoxin-based drugs suffer from having a narrow therapeutic range, meaning there is a fine line separating an effective dose from one that is toxic. Vincent van Gogh is thought to have suffered from the toxic effects of digitoxin, one of which is the perception of a pervasive yellow-green tint. Van Gogh's penchant for yellow is clearly visible in his *Portrait of Dr. Gachet*, a painting of his physician with a stem of purple foxglove. Perhaps van Gogh's physician was prescribing too much digitoxin, or another possibility is that van Gogh's microbiota was missing a particular gut microbe, one that could have protected him from the drug's side effects.

The bacterium *Eggerthella lenta*, a resident of the microbiota of some individuals, contains a suite of genes that can deactivate the drug digoxin. So whether or not your microbiome has this ability could affect the correct dosage of digoxin. A person housing *Eggerthella lenta* would presumably need a higher dose than someone without that microbe. However, digoxin and the amino acid arginine are consumed by *Eggerthella lenta* using a common pathway. Mice with *Eggerthella lenta* eating a high protein diet couldn't deactivate digoxin very efficiently since the bacteria were so busy consuming arginine that there was no capacity left to work on digoxin. In this case both the composition of the microbiota and the protein content of the diet influence the dose of digoxin that a patient would experience.

Let's take a moment to think about what this means for the future of personalized medicine and how our knowledge of the microbiota will influence medical treatment. You can imagine a scenario in which an individual would benefit from taking digoxin. Before prescribing this medication, a physician would obtain a report listing the gene content of the patient's microbiome. If the patient's microbiome had the digoxin-deactivating genes, the doctor would prescribe a higher dose of the drug along with instructions to minimize his or her consumption of protein. The result would be a precise, individualized digoxin dose, one that was effective yet with the threat of side effects reduced.

THE "FOUNTAIN OF YOUTH" IS FILLED WITH BACTERIA

Age-related health decline is tied to a variety of outward physiological manifestations such as decreased mobility, loss of mental acuity, and loss of vision and hearing. But inside our bodies, aging results in immunosenescence, inflammaging, and changes to the microbial community in the gut. There may be no escaping the pitfalls of aging, but there are ways to delay or slow the pace of health-related decline.

The microbiota, inseparably linked to human metabolism and the immune system, impacts aging. It seems that taking advantage of the malleability of the microbiota to minimize the damage and deterioration it incurs as we age could improve the quality of our later years. Perhaps a well-maintained microbiota could infuse new life into the senescing immune system, thereby keeping our immune system and our body younger, longer.

Human examples of extreme longevity, centenarians, have a microbiota profile that is distinct from that of seventy-year-olds. Is there something special about the centenarian microbiota that contributes to extreme longevity? Or do genetics or lifestyle choices of these persevering individuals foster a certain microbial community? We don't yet know the answers to these questions. But perhaps the secret to a long life is when human cells and microbes, joined in an optimal symbiosis, nudge each other onward in a mutually beneficial, long-lasting interaction.

KEEPING THE MICROBIOTA YOUNG

There are many ways to maintain some youthfulness over time. Eating a highly nutritious, well-balanced diet, getting adequate exercise, and having a robust social network all have been scientifically shown to improve health as we age. How these behaviors contribute to better aging are likely multifactorial, affecting many aspects of our biology ranging from maintaining muscle mass to providing a sense of purpose to our lives. But as scientists dissect the molecular mechanisms to explain exactly how these healthy habits prolong our lives, the microbiota will emerge as an important player in determining how well we age.

A major part of the benefits of a healthy diet is its ability to promote a flourishing microbiota and the ensuing cascade of health effects. The study of the Irish seniors demonstrates how maintaining a diet rich in fiber (or MACs) and low in fat can protect the microbiota from age-related decline. In fact, increased fiber consumption among seventy-six- to ninety-five-year-olds is directly related to elevated SCFA production. The role that SCFAs play in minimizing inflammation could be an important counterbalance to the ill effects of inflammaging. Maintaining a highly nutritious diet is especially critical as we age because calorie requirements decline over time. Each calorie really has to pull its weight. Scientists at Tufts University have publicized a modified version of the USDA's MyPlate dietary recommendations specifically for older adults to adjust for the lower calorie requirements. In their rendition, there is a focus on nutrient-dense, brightly colored fruits and vegetables, and fiber-rich whole grains and beans are prominently featured.

Probiotics offer another dietary avenue to improve the aging microbiota. Research investigating the effect of probiotics in older individuals is likely to surge considering the increasing age of the U.S. population. Current findings do suggest that probiotics can help the aging immune system. As new probiotic products are developed it will be important to keep in mind that the aging microbiota is distinct from that of the young or adult microbiota. Probiotic bacteria with a special set of functionalities that can thrive in this altered environment will be required to properly complement the aging microbiota. The age of a one-size-fits-all model of probiotics is passing. Future probiotic therapies are likely to account for the development of the human microbiota throughout the stages of life. Until such age- and individual-specific probiotics are available, it's up to us, the consumers, to use a trial-and-error method to identify products that agree with each of us.

Exercise, while clearly able to improve health, is an unknown quantity with regard to its impact on the microbiota. Studying the effect of exercise on the human microbiota is complicated because people who exercise are more likely to have a healthier diet.

Distinguishing what microbiota difference is driven by exercise as opposed to diet is nearly impossible using human subjects. However, laboratory animals that exercise have differences in their microbiota compared to that of sedentary animals when fed the same diet. Several physiological changes that result from exercise, such as increasing intestinal transit time (or flow rate) through the gut, influencing metabolism, and altering immune function, are known to affect the microbiota. So it's not much of a stretch to assume that exercise, through its ability to alter so much of host biology, will change the microbiota in humans. While the jury is still out on whether physical activity has positive effects on the microbiota, exercising may improve microbiota health and its positive impact on overall health is a safe bet.

The link between a robust social network and good health as we age, while still an emerging branch of science, is clearly important. How the microbiota fits in here is entirely speculative at this point, but let us provide this food for thought. Remember that microbes, including those from our gut, are everywhere. Try as we might to completely sterilize our environment, it is almost impossible to remove the patina of human-associated microbes that cover many of the surfaces we touch (although many would argue that this patina is slowly being chemically stripped away). This means that through contact with other people while playing cards at the local community center, sharing a meal with a friend, or shaking hands with others at religious services, you are also encountering their associated microbes. Is it possible that part of the antiaging effect provided by social connections could be coming from contact with other microbes? Before completely disregarding this idea as something only a microbiota-obsessed scientist would come up with, ponder this: researchers found that cohousing lean mice with obese mice allowed "lean" microbes to infiltrate the microbiota of the obese mice and protected them from weight gain. If the obese mice had not had contact with the lean mice they would not have been exposed to the "lean" microbes and would be stuck with their "obese" microbiota.

It is possible that by exposing ourselves to other people's microbes through social contact we could be inoculating our gut with a diverse array of beneficial microbes. Perhaps this is just a crazy idea, but ten years ago so was the idea that gut microbes could contribute to obesity. Undoubtedly there are many benefits to maintaining social networks as we age. But the next time you are contemplating whether to join the local bridge club, remember that some of the new friends you acquire there might be microscopic lifesavers.

CHAPTER 9

Managing Your Internal Fermentation

YOUR GENOME IS NOT YOUR DESTINY

While there is nothing we can do to change our human genome, our microbiome offers opportunities to exert control over the genetic hand that we've been dealt—akin to asking for a couple of replacement cards in draw poker. Changes to our microbiome can't affect our eye color or the shape of our nose, but many aspects of our biology, such as our weight and immune system, are heavily influenced by our gut microbes.

One could argue that the composition of our microbiota could be predestined, in a way, by our human genome. Our human genes create the gut environment that these microbes inhabit. Perhaps which microbes take up residence is largely a product of certain human genes we inherited at birth, making the composition of one's microbiota a fait accompli. If that were the case, then identical twins should have a microbiota that is more similar to each other's than those of fraternal twins. But in fact this is not the case. Environment plays a huge role in our internal collection of bacteria. Since there is much we can do to shape the environment within our gut, we have control over our microbiota and can compensate for the lack of control we have over our human genome. Our microbiome contains one hundred times more genes than our human genome, so in fact there is about 99 percent of associated genetic material that we have the potential to mold in ways that are beneficial to us.

Knowing that our microbiota is adaptable is not enough. We need to learn how to exact positive change on our microbiome to maximize the benefit it can provide to our health. The following section outlines specific habits we can foster to both optimize the community and our health. We have incorporated all of the following recommendations into our own and our children's lives. Each recommendation is rooted in the scientific work performed by our lab and those of others in the field during this past decade of unprecedented microbiota enlightenment.

GETTING A JUMP START ON A HEALTHY MICROBIOTA

During infancy the microbial equivalent of a land grab is occurring in the gut. Species that are successful in establishing early can persist for decades, and perhaps throughout life. There are many factors that affect which microbes colonize, among them mode of delivery, diet, antibiotic use, and exposure to environmental microbes. Nurturing the microbiota at the beginning of life can help get this symbiotic relationship started out on the right foot.

The way a child is delivered is the product of a multitude of factors, many of which are out of our control. Clearly the safety of both the mother and her newborn is the highest priority, but how the numerous choices that confront expecting parents influence the microbiota is worth considering. From a microbiota standpoint, a vaginal delivery exposes the infant to a set of bacteria very different from the skin microbes that colonize a C-section–delivered baby. The composition of the mother's microbiota changes during her pregnancy, presumably to better care for the growing fetus and to seed the newborn with the most beneficial "starter" microbes. But having a C-section delivery does not rule out exposing the infant to the microbes he or she would have received through a vaginal birth. Discussing with your physician the possibility of inoculating a C-section baby with a vaginal swab after delivery could help put the early microbiota on the trajectory that nature intended.

One of the largest levers we have to control the inhabitants in the gut is diet. The first foods that babies experience influence which microbes dominate the gut during immune system development and education. Until there is a complete understanding about which collection of microbes predisposes one to or protects one from allergies, asthma, or even obesity, the safest bet is to provide nourishment that has stood the test of time. Breast milk has been engineered by the forces of evolution, working to maximize our species' fitness over thousands of generations. This optimal infant food contrasts to formula, which is a result of only decades of human tinkering. Breast milk, and specifically the quintessential microbiota accessible carbohydrate (MAC) human milk oligosaccharides, is a microbiota superfood. While infant formulas containing prebiotics and even probiotics are available, none have been optimized the way that breast milk has. The reality is that because we don't understand the complexity of factors that impact the developing microbiota, science has not yet designed or manufactured a formula that guides microbiota development the way breast milk does. Breast milk offers the infant an opportunity to start life on a high-MAC diet.

Like the method of delivery, how we choose to feed our infant is often guided by circumstances that are beyond our control. But breast-feeding does not have to be all or nothing. Any amount of breast milk will provide the infant with human milk oligosaccharides and the microbes found in breast milk that are not currently available in any other form. Nightly nursing before bedtime can be a great way to calm both mom and baby at the end of the day and provide sustenance for the baby's developing microbiota. We urge mothers who are having difficulty breast-feeding or feel they are not producing enough milk to seek outside help through a support group or lactation consultant. Breast-feeding is a complex skill that requires time and practice to master. Although admittedly a lot of work, the effort you invest to establish breast-feeding with your child will be rewarded with a lower risk of allergies, asthma, obesity, and even diabetes. Your child, his or her microbiota, and subsequent generations that inherit this microbiota will be eternally grateful!

THE PROBLEM WITH ERADICATION

Antibiotics are a marvel of modern medicine. They have saved countless lives and continue to be one of the most effective classes of drugs ever created. But their effectiveness is also what makes them so potentially dangerous. Most antibiotics don't discriminate between the "bad" bacteria they are tasked with eliminating and the "good" ones that get caught in the crossfire. This collateral damage can get more and more difficult to repair, with each antibiotic exposure taking a toll on our microbiota's diversity and opening up our risk to opportunistic infections from pathogens like *C. difficile* and *Salmonella*.

There are times in our life when taking antibiotics is unavoidable, but it is clear that as a society we are overusing them, especially when it comes to children. The one course of antibiotics per year that the average American child is prescribed could permanently change that child's microbiota and affect his or her long-term health. If we want to protect the health of the microbiota as well as minimize the prevalence of antibiotic-resistant superbugs, antibiotics need to be used sparingly, and then only when absolutely necessary. Each time our family is faced with the possibility of using antibiotics, we perform a cost-benefit analysis with the help of our physician. If the doctor feels it is safe to take a wait-and-see approach, we do. If the doctor decides that antibiotics are the best course of action, we use them. In the two cases when our children were prescribed antibiotics, we gave them probiotics, once in supplement form for a newborn, and in the second instance with yogurt, during and after the antibiotic treatment.

To create an environment where antibiotic use is infrequent, prevention is key. Having school-age children means that our household is a revolving door of runny noses and scratchy throats. Eating a nutritious diet and probiotic foods can help limit the number of illnesses and minimize their duration. A healthy diet coupled with plenty of sleep can synergize to help keep sickness at bay. Flu season is one of the few times that we are vigilant about hand-washing, typically making our kids reach for the soap after they come home from school to curtail the spread of infectious microbes (mainly viruses) from classmates. But as important as limiting exposure to pathogens is supplementing the body with beneficial bacteria. Starting each day with a glass of kefir or bowl of yogurt can provide billions of microbes to help boost your body's defenses.

INCREASING OUR MICROBIOTA'S SOCIAL NETWORK

The modern Western lifestyle has resulted in an alien environment for the gut microbiota. Our overly sterile surroundings make encounters with microbes far less numerous than they were when floors were made of earth and washing produce meant brushing dirt away with your hand. But there are ways to provide more company to our resident inhabitants without forgoing the benefits that modern-day sanitation provides. Animals provide added microbes to counterbalance our ultraclean existence. Homes on farms contain a much higher diversity of microbes than urban dwellings. Children raised on farms are much less likely to suffer from asthma and allergies, presumably because of increased encounters with environmental microbes.

While moving to a farm is not practical for most of us, there are still ways to surround yourself with more microbes in an urban setting. A small garden can be a conduit to increased microbial interactions. If space for a garden is limited, explore creative ways to use the space that you have. Pots on a patio or even a window-box herb garden can encourage contact with natural microbial life that occurs in soil and on plants. With space being a premium in the San Francisco Bay Area, we have converted a portion of our front yard to raised-bed garden boxes. Our children love to dig their fingers into the dirt to pull weeds, play with worms and grubs, or harvest whatever vegetables happen to be ready. Because we don't use herbicides, pesticides, or synthetic fertilizers, we feel comfortable letting our kids go from the garden to lunch without washing hands first. If you don't have space for a garden, visiting an organic farm is not only an educational outing but also a great opportunity to introduce your microbiota to some new "friends." Many farms that sell produce directly to consumers in the form of community supported agriculture programs, or CSAs, allow members to tour their farms. Some of these farms will even allow city slickers to spend a few hours of the weekend pulling weeds and harvesting crops. These experiences can provide exposure to environmental microbes that may not be present in an urban or suburban home.

Children who grow up with family pets are, like children raised on farms, protected from respiratory infections and allergies and are less likely to require antibiotics compared to children in pet-free homes. Pets can bring beneficial microbes they collect from outside or that reside on them into our homes. We have routinely witnessed our dog sniff around our backyard, nuzzle in the dirt, and then run over to give our children a great big lick on the face—a vivid illustration of the increased microbe exposure a pet can provide. (Dirt is just one example of many microbial-rich environments that our dog explores.) Petting the family dog is not an automatic hand-washing trigger in our home. Our dog is free from topical flea medication, is routinely checked for transmissible intestinal parasites, and spends most of his time in our pesticide- and herbicide-free yard. In this situation, we feel the potential benefit of being exposed to more microbes outweighs the risk of harm from not thoroughly washing hands.

Pet-averse people need not worry; humans can also be a source of "extra" microbes for children. A recent study found that children whose parents cleaned their pacifier by sucking on it as opposed to rinsing or boiling it in water were much less likely to have eczema, a condition in which skin becomes inflamed or irritated. Children using the mommy-cleaned pacifiers were no more likely to develop a respiratory infection than those with more vigorously cleaned pacifiers. This study provides a nice example of how relaxed cleaning can result in a better health outcome with no apparent downside. Might this principle apply not only to pacifiers but also to how we clean our homes? Don't worry, we're not advocating licking your entire home to clean it. But maybe using antibacterial household cleaners or bleach is equivalent to boiling pacifiers as far as our health is concerned. A more microbe-friendly approach to cleaning is to use less-toxic cleaners such as vinegar, castile soap, and lemon juice, which will allow increased exposure to microbes and may lessen the risk of the misfiring immune system that is plaguing the Western world.

EATING FOR YOUR MICROBES

The main goal of the dietary interventions we recommend is to improve the diversity of bacteria within your microbiota and increase the amount of SCFAs produced through bacterial fermentation. Multiple scientific studies show that people with a diverse microbiota producing lots of SCFAs are healthier and less prone to Western diseases than people with fewer bacterial types in their gut. People consuming a diet and living a lifestyle more similar to our ancient ancestors' house a more diverse microbial community than Westerners. Even within the Western population, lean people harbor greater bacterial diversity than obese people. Overweight and obese individuals with the lowest diversity also have the highest incidence of insulin resistance, elevated cholesterol, and inflammation compared with obese people with a more varied microbiota. Overwhelming evidence is building for a connection between better health and a complex microbiota. How can a high diversity microbiota be encouraged? By creating an environment within the gut that welcomes and sustains many different types of bacteria, an environment replete with sustenance in the form of dietary MACs.

The good news is that since the microbiota is so responsive to dietary changes, making wise food choices can be an extremely effective way to improve the state of your microbiota. It is important to realize that microbiota responsiveness works on short and long time scales and can change in different directions, depending upon the change in diet. Increasing your dietary MACs will lead to a rapid shift of the microbiota, but removing dietary MACs will also have a rapid shift in a different, and likely less healthy direction. The key is to increase nourishment to the microbiota and maintain this MAC-rich diet over time. Long-term dietary patterns are a major determinant of achieving and maintaining microbiota diversity. Our recommendations are made in the spirit of a fundamental rethinking about the way we eat, a way that considers nurturing the microbiota. There is no microbiota crash diet, because although the microbiota responds quickly to dietary change, it is long-term dietary patterns that translate into the lifelong positive health effects provided by the microbiota.

There are four main tenets of a microbiota-friendly diet. The first is to consume foods that are rich in dietary MACs. The bacteria in our gut need food, and the cuisine they covet most is carbohydrates. There are two main sources for these carbohydrates—from the MACs found in the dietary fiber we eat and from the protective layer of mucus lining the gut. Ideally, most of the carbohydrates for bacterial consumption would come from dietary inputs and not from the intestinal mucus lining, which provides a functional barrier that keeps bacteria at a safe distance, a border fence of sorts. By encouraging microbes to associate tightly with this barrier to obtain nourishment (say by depriving bacteria of dietary carbohydrates) we risk enriching for a gut community that specializes in eating mucus, which could compromise the protective layer.

When deciding what to put on our plate, we need to be mindful about what is going to feed the microbes waiting at the end of our digestive system. A breakfast of eggs, bacon, white-bread toast, and pulp-free orange juice is going to provide almost no MACs to sustain your microbes and no fuel for SCFA production. If that breakfast is followed by a lunch consisting of a white-bread sandwich, chips, and a soda, now your microbes have missed two meals. Finish off the day with a piece of meat, mashed potatoes, and a couple of limp, overcooked florets of broccoli and your microbiota has had an entire day of minimal dietary MACs. Under these circumstances, the microbiota hits the only food reserve it has—you. It consumes carbohydrates found in your intestinal mucus, inching ever closer to the lining of your gut and compromising the barrier your body has constructed to keep microbes at a safe distance. If this same scenario plays out day after day, there is a risk that your immune system will become alarmed and retaliate in the form of colonic inflammation.

Keeping your microbes away from the mucus lining is just one reason to consume more dietary MACs. The other is that a diet rich in MACs can help sustain a much more diverse population of microbes. If the same intestinal mucus carbohydrates are always on the menu in the gut, it will limit the collection of microbes that can thrive. But if you provide a varied assortment of carbohydrates from a diet rich in fruits, vegetables, and grains, suddenly the carbohydrate choices are immense and many types of microbes can bloom. This can lead to a stable community, one that provides robust protection from invading pathogens and is equipped to produce health-promoting SCFAs. Obese and overweight individuals on a low-calorie, high-fiber diet (containing 30 percent more dietary fiber and 130 times more soluble fiber than their previous diet) lose weight and gain microbiota diversity. Along with the increased diversity comes a lower risk for diabetes, atherosclerosis, and cancer. By increasing the amount of dietary fruits and vegetables—and the MACs they contain—these people created an environment within their gut that allowed their microbiota to flourish. In return they were rewarded with indicators of better health.

A second important facet of a microbiota-friendly diet is to consume meat in limited quantities. Red meat contains the chemical L-carnitine, which certain microbes in the gut can convert to trimethylamine (TMA), which then gets oxidized into trimethylamine-*N*-oxide (TMAO). Regular meat-eaters have more TMAO than vegans or vegetarians. High levels of TMAO increase the risk of strokes, heart attacks, and other cardiac events. Long-term dietary patterns impact the ability of the microbiota to produce this dangerous compound. People consuming a plant-based diet with little to no meat produce less TMAO when they do eat meat, likely because they don't have as many TMA-producing bacteria in their gut. Ideally, it would be possible to determine whether your microbiota contains many TMA-producing microbes so that dietary choices could be made accordingly. Unfortunately, we don't yet know enough to be able to make these types of predictions. Until it is clear which microbiota compositions are resistant to TMAO production regardless of diet, it seems safest to limit the consumption of meat, especially red meat, which has the highest levels of L-carnitine.

The third pillar of a microbiota-friendly diet is limiting saturated fat intake. Diets high in saturated animal fat are detrimental to microbiota diversity. The bacteria that are able to flourish on a high-fat diet include those mentioned in the previous chapter, the pathobionts, residents of the microbiota that can trigger inflammation in the gut. Plant-derived monounsaturated fats don't promote pathobionts as readily. Biasing your fat intake toward foods like olive oil or avocados will satisfy your cravings for fat without bolstering the pathobionts lurking in the gut.

The final component of a microbiota-friendly diet is consuming beneficial microbes, or probiotics. Humans have a long history of ingesting bacteria living on or in our foods. Before refrigeration and sanitation became widespread, eating bacteria on spoiling, unwashed food was part of everyday life. Today, consuming bacteria like those found in fermented foods such as yogurt reduces the risk of illness from food-borne and respiratory pathogens. However, relaxed oversight of probiotics by the FDA, combined with overly exuberant claims of what these products can do, has led to a market that is confusing for consumers to navigate. To add to the complexity, a one-size-fits-all model does not apply to probiotics. Therefore, finding a probiotic that is beneficial for you requires that you become a microbiota experimentalist. It is best to test a variety of probiotics to determine what works for you.

If you have a specific health concern, you should consult your doctor to determine whether a particular probiotic product is appropriate. Our family consumes bacteria primarily in the form of fermented dairy products like unsweetened yogurt and kefir but we also enjoy fermented vegetables such as pickles and sauerkraut. These are just our personal preferences. There is a whole world of fermented foods available, some of which we have listed in the Appendix. This list is not meant to be exhaustive since the number of products containing microbes is increasing as public awareness grows, but may help provide a starting point for those new to the world of fermented foods. Try a variety of fermented foods to see what suits you and your microbiota best and begin incorporating them into your everyday cuisine.

The type of probiotic that is the best fit for you is likely to be as individual as your microbiota. Therefore it may take some experimenting with different probiotics to find something that agrees with you. Any probiotics that cause painful bloating or digestive distress are not playing nice with your gut or microbiota. Ideally, probiotics should promote regular bowel movements that are easy to pass. However, it may take some time for probiotics to exact a perceptible positive change, so some patience is required.

There are many types of probiotics to try both from fermented foods and from supplements. A systematic approach that involves trying a particular probiotic, say a certain brand of yogurt, daily for at least a week will help you determine whether that product is providing a benefit. If you feel that it's not, move on to a different type of fermented food and try again for at least a week. If you prefer to take supplements, go with manufacturers that you recognize and who have reputations at stake, since they are more likely to have good quality control. You may want to try probiotic supplements that contain multiple microbial strains, which more closely mimic the diversity of microbes found in fermented food. As you try different products, beware of highly processed food technology (often cloaked in highly decorative and colorful packaging) that attempts to masquerade as health food by adding in some probiotic cultures. Consuming bacteria-laden foods will expose the microbiota to a steady stream of environmental microbes in a way that approximates our ancestors' diet.

There are times when it may be worth loading up on beneficial bacteria. If at our house we experience a bout of food poisoning or the so-called twenty-four-hour flu or feel like we might be coming down with a cold or sore throat, we increase the amount of microbes in our diet, usually in the form of an extra glass of kefir daily. Another time to boost microbe consumption, of course, is following a round of antibiotics.

We have provided information about how our family uses probiotics to serve as an example of how probiotics can be incorporated into your life. There is essentially no evidence that any particular probiotic strain is more beneficial than any other, only that, in general, the consumption of fermented food–associated (or food-derived) bacteria is healthful. Keep in mind, however, that many probiotic foods may be laden with added sugar, especially those marketed to children. Choose products with the fewest ingredients possible and with little to no added sugar. Ideally, the ingredient list should contain more bacterial species than unrecognizable ingredients. And if cane sugar, corn syrup, or any other type of sugary sweetener is listed within the first three ingredients, avoid the product! If your children have a difficult time with the sour, acidic flavor that dominates unsweetened fermented foods, consider adding small amounts of honey or maple syrup, then gradually reducing them. With small reductions in added sweetener over time, you can help your children become accustomed to enjoying unsweetened fermented food.

As you may have noticed, this microbiota-focused diet shares many features with the Mediterranean and the traditional Japanese diets. These diets support exceptional health and longevity. Likely not by coincidence, these diets have in common high fiber content, low saturated fat, low red-meat consumption, and the regular incorporation of fermented food. While the ways in which these diets support health are undoubtedly complex and multifactorial, we are now beginning to understand that one important facet involves supporting a healthy microbiota.

A MICROBIOTA-FRIENDLY DIET IN PRACTICE

Knowing that the microbiota thrives on a diet replete with MACs, low in the consumption of meat and saturated fats, and with plenty of probiotic bacteria, how can a microbiota-friendly diet be achieved in practice? The Appendix provides a sample week of meals that nourish the microbiota. Each day contains 33 to 39 grams of dietary fiber from different sources to maximize the diversity of MACs your microbiota can ferment. This amount is based on the recommendation from the USDA dietary guidelines that 14 grams of fiber should be consumed per 1,000 calories consumed. A more personalized amount of fiber recommended, based on gender and age, by the Institute of Medicine of the National Academies is also provided in the Appendix. Each day contains at least one probiotic food item to supplement your resident microbes with some environmental visitors. Most days are meat free to limit potential TMAO production, and fat is primarily derived from plant sources to nudge the microbiota toward a community more focused on plant-based MAC fermentation (and increased SCFA production). Recipes are provided for many of the dishes listed. This one-week meal plan should serve as a guide to help you create microbiota-friendly menus each week.

Having school-age children, we understand the struggles that accompany packing healthy, nutritious school lunches. We have included school lunch ideas that we routinely use for our children's lunches. Unless your children attend a school that is highly committed to providing healthy lunches—and such schools seem few and far between—packing your child's lunch will ensure that they are exposed to nutritious, microbiota-nourishing food and allow you to keep track of what your child is (or is not) eating. The cafeteria at our children's school provides a salad bar loaded with fiber-filled fruits and vegetables, but unless someone places these foods on their plates and reminds them to eat their "growing food," most kids will skip these items. Even as adults who understand the repercussions of our food choices, it can be hard to make healthy choices. How can we expect our kids to select and eat a plate of raw veggies over the alternative cafeteria options such as a cheeseburger and fries or cheese pizza? We applaud the efforts made by schools to provide kids with healthier lunch options, but the unfortunate truth is that without reinforcement at home, kids often won't make the healthy choice. To bring about lasting change to the microbiota, eating in a microbiota-friendly way needs to be a long-term commitment.

One final note on the elephant in the room, flatulence. Often, switching to a diet rich in MACs can lead to a short-term increase in gas production. However, over time your microbiota will adjust and gas production will normalize. Many people we've talked to about the importance of a high-fiber diet complain about the uncomfortable bloating and gas production they experience and use it as a reason to lower their fiber consumption. To minimize discomfort, it may help to increase dietary fiber slowly to give your internal fermentation a chance to adapt to the increase in microbiota accessible carbohydrates. By gradually increasing dietary fiber you can eventually achieve the recommended amount of fiber in a way that minimizes discomfort. Once you reach this optimal level of fiber, it is important to continue eating plenty of fiber to maintain homeostasis. The pace at which you are able to add fiber to your diet will depend on a number of factors, for example, how much fiber you've eaten historically and the personal nature of your microbiota. Pay attention to how your body is responding to this dietary change and go as slowly as you need to (or as quickly as you can tolerate). Just maintain a trajectory that will get you to 25 to 38 grams of fiber per day. It may take a couple weeks, or even months, to get there, but eventually your microbiota will adjust to this new way of eating. The best impact you can have on your microbiota is to attain a sustainable, long-term microbiota-friendly diet, so be patient but also strive for a goal. As the Inuit experienced each year with the reintroduction of dietary MACs, the process of increasing nourishment to the microbiota can be a bit uncomfortable. But unlike the Inuit, our access to MACs is not limited by the seasons, so taking a gradual approach can minimize discomfort as your microbiota adjusts. Taking a scientific approach to increasing MAC consumption can help you identify good sources of MACs and probiotics that are the most compatible with your microbiota and digestive system. Some people experience food sensitivities, which can manifest in a wide range of symptoms, including bloating, gas, headaches, and lethargy. For example if you notice a sensitivity to gluten, try other types of gluten-free grains such as quinoa, millet, or buckwheat. Injera, a fermented flat bread made from teff flour, is high in fiber, gluten-free, and contains compounds from microbial fermentation (however, the microbes are killed during cooking). How legumes are tolerated can also be highly individual. If you find that garbanzo beans are resulting in too much discomfort, try black beans or lentils.

We encourage you to follow the changes occurring within your microbiota by participating in the American Gut Project. Although we are not involved in this crowd-funded science project, it is run by a team of well-respected scientists and has provided thousands of people with information about their microbiota. You can have your gut microbiota sequenced before and during your process of microbiota improvement to witness the changes to the new aspects of your diet and lifestyle. You will be provided with a report specifying the types of microbes that make up your microbiota and how it compares with others who have participated as well as to people living in developing regions of the world (Malawi and Venezuela). This information will not only allow a better view of your microbiota and how it compares with others, but will also contribute to the scientific understanding of these communities. To guide you in your journey of microbiota revitalization, we recommend submitting multiple samples—an initial sample to document where your microbiota started out, then one or more after you have made dietary and lifestyle adjustments in order to see how these changes are impacting your gut community over time. This will not only be informative but may also motivate you to keep improving the health of your microbiota.

BEYOND THE COLLECTION OF BACTERIA IN OUR GUT

By now you have a much better understanding of the collection of bacteria that call our gut home and influence our biology in numerous, interconnected ways. The gut provides shelter to the largest collection of bacteria we carry, but many other sites on our bodies are colonized. Our mouths, skin, noses, lungs, ears, vaginas, and even our belly buttons are habitats for microbes. All the microorganisms inhabiting these varied locations are an integral part of the superorganism we call the human being. And while research on these other communities currently lags gut microbiota research, all of them play a role in our health.

Our microbiota is experiencing a change in its habitat that it has not seen since the birth of agriculture more than ten thousand years ago. The modern Western way of eating, with minimal MACs and limited consumption of microbes, coupled with the rising use of antibiotics and antibacterial products, presents numerous challenges to the microbiota. These challenges have resulted in a gut community that is less diverse and appears to be missing key species compared to that of modern individuals living a more traditional lifestyle—the same individuals who enjoy a lower risk of Western diseases. Luckily, the same plasticity that has allowed the Western microbiota to deviate so quickly from that of our ancestors can also help fuel its rebirth. By nurturing your gut microbes through improving diet, minimizing antibiotic use, and reconnecting with nature (and all the microbes it contains) you can improve the health of this community.

As the intricate, interwoven, multispecies interactions that define our biology emerge, we need to adopt a new definition of what we are as humans. This definition should take into account the expanse of organisms that forms our collective mosaic of cells. We are a composite organism, an ecosystem. As we think about our health we need to be mindful of the microorganisms we harbor and think about what effect our diet, lifestyle, and medical decisions will have on our microbial self.

MENUS and RECIPES

MICROBIOTA-FRIENDLY 7-DAY MENU

Approximate amounts of dietary fiber are included for reference. An asterisk denotes menu items for which a recipe is provided.

Sunday—34.5 grams of fiber

BREAKFAST (17 grams of fiber)

Symbiotic Scramble *

Aztec Hot Chocolate *

LUNCH (5 grams of fiber)

Salade Niçoise

SNACK (3.5 grams of fiber)

Fermented-Filling Dates *

DINNER (9 grams of fiber)

Kale Pesto with Whole Wheat Pasta

Figs

Monday—35.5 grams of fiber

BREAKFAST (8 grams of fiber)

Bacteria-Boosting Granola *

Blueberries

LUNCH (16 grams of fiber)

Chickpea Greek Salad *

SNACK (2.5 grams of fiber)

Japanese Popcorn *

DINNER (9 grams of fiber)

Fiber-Filled Flatbread Pizza *

Tuesday—39.5 grams of fiber

BREAKFAST (7.5 grams of fiber)

Muesli for Microbes *

LUNCH (19 grams of fiber)

Kale Salad with Chia Seeds, Pomegranate Seeds, and Pistachios

SNACK (3 grams of fiber)

Cashews for Your Commensals *

DINNER (10 grams of fiber)

Sausage, Onions, Potatoes, and Sauerkraut

Cup of Raspberries

Wednesday—36 grams of fiber

BREAKFAST (9 grams of fiber)

Middle Eastern Oatmeal Pudding *

LUNCH (6 grams of fiber)

Sandwich on Whole Wheat Bread with Fermented Cream Cheese, Smoked Salmon, Canned Artichoke Hearts, Tomato Slices, and Capers

DINNER (21 grams of fiber)

Microbiota-Reframed Risotto *

1 Ounce Dark Chocolate

Thursday—33 grams of fiber

BREAKFAST (10 grams of fiber)

Whole Grain Toast

Almond/Walnut Butter * and Strawberry Slices
Morning Microbiota Smoothie *

LUNCH (7 grams of fiber)
Totally MAC-Filled Tabbouleh *

SNACK (3 grams of fiber)
Banana

DINNER (13 grams of fiber)
Sesame Seed–Crusted Salmon with Green Beans and Orange Miso Sauce * and Brown Rice

Friday—34 grams of fiber

BREAKFAST (7 grams of fiber)
Crunchy Yogurt Parfait *

LUNCH (11 grams of fiber)
Soba Noodle Salad with Probiotic Peanut Miso Sauce *

SNACK (8 grams of fiber)
Hunter-Gatherer Tuber Snack *

DINNER (8 grams of fiber)
Mutualist Mediterranean Soup *

Saturday—36 grams fiber

BREAKFAST (9 grams of fiber)
Tarahumara Pancakes *

LUNCH (10 grams of fiber)
Multigrain Crisp Bread Layered with Spinach, Sardines, Red Pepper Slices, and Chives, with a Squeeze of Fresh Lemon Juice
½ Cup Blackberries

SNACK (4 grams of fiber)
Apple

DINNER (13 grams of fiber)
Indian Dal* Served over Brown Rice
Mango Kefir Lassi*

FEEDING YOUR MICROBIOTA

We hope that these recipes will provide a window into the type of dishes that help sustain the microbiota. They are a regular part of our diet, and we have tried to make them as simple as possible to prepare and manageable for weekdays. Because dietary fiber is a good indicator of MAC content and can be tracked using existing nutritional information, we've listed the dietary fiber content for the recipes rather than the microbiota accessible carbohydrates (MACs).

These recipes contain minimal amounts of sugar. Over the past few years we have consciously reduced our intake of sugar, preferring to consume carbohydrates in a form that benefits our microbes. You may find that some of these dishes are not as sweet as you are accustomed to. Limiting sugar in your diet requires that your palate readjusts. At first you may need to include more sweetness, especially if you are trying to coax children into eating healthier. But over time the added sugar can be reduced and your dietary fiber intake increased. Eventually you may find that very little sugar is needed to satisfy your sweet tooth—and that baked treats are too sweet to enjoy and traditional recipes need adjustment. Meanwhile, the additional dietary fiber will keep you feeling satisfied longer without craving simple carbohydrate-loaded empty calories.

Almost none of these recipes are the kind you would find on the kids' menu at a typical restaurant. Other than the school lunches, there are no dishes that are kid specific. This is intentional. Most traditional "kids'" foods contain little to no dietary fiber, often feature copious amounts of cheese and/or processed meats, and are disastrous for the microbiota. It may be a challenge to get some children interested in more-healthy food choices. We can offer a few tips that have helped our children learn to love microbiota-friendly meals.

First, do not be discouraged when kids reject a new food item. Often it can take many introductions (sometimes ten or more) for kids to accept and eventually enjoy a variety of legumes and vegetables. In our experience, persistence is quite effective in shaping healthy eating habits. Of course, modeling healthy eating is critical: it is important that your children see you enjoying healthy food as well. Mentioning how happy your microbiota will be with better food doesn't hurt either. Gardening and preparing meals with your children can help motivate them to try new dishes. Finally, it helps to remember the importance of a microbiota-sustaining diet for everyone's health. Approach mealtimes with the same resolution and firmness you would with other activities that are critical to health and well-being. Just as you wouldn't yield to your child's requests to skip a day of school or routinely stay up too late, starving the microbiota should be viewed as unacceptable. Knowing that you are providing healthy food for your family (and their microbiota) and staying aware of the lasting benefit this can provide throughout their lives can help reinforce your resolve when your child screams *yuk!* at the sight of the nutritious meal you served.

A final couple of notes about our cooking style. We typically have our food processor and blender either on the counter or in a cabinet that's easy to access. Several recipes rely on these two appliances to chop or puree ingredients. In cases where beans are a part of the recipe, we recommend that they be cooked at home as opposed to coming from a can. This requires a little more planning, but not too much effort; and there's a great reward in taste. We usually cook a pot of one type of bean over the weekend (a couple hours on a simmer is enough for most types of dried beans). Then we store the cooked and drained beans in mason jars that we either freeze or keep in the refrigerator. This advance preparation enables us to throw beans into a mixed green salad with nuts and seeds, or into a soup for a fiber-rich dinner. Of course, canned beans can easily be substituted and do save time in a pinch.

Breakfast FOR YOUR GUT BUGS

Conventional wisdom (and mounting scientific evidence) holds that breakfast is the most important meal of the day. But much of what we consume for breakfast is not the best for our health or our microbiota. Breakfast items often seem to fall into one of two traps. One is the white-flour-based pastries, pancakes, and the like that are usually loaded with sugar or drenched in syrup. The other trap is the breakfast filled with animal products like eggs and bacon and paired with a slice of buttered white-bread toast. Neither of these options offers much in the way of microbiota accessible carbohydrates. Imagine you're a member of the microbiota waiting to break your fast in the morning only to have to wait for lunchtime, when dietary fiber often makes its first appearance of the day. Here are a few ideas for getting your microbiota off to a good start in the morning.

Morning Microbiota Smoothie

SERVES 2

(3.5–6 grams of fiber per serving, depending on fruit and greens choice)

It can be difficult to get enough fresh vegetables in our diet when the first meal of the day rarely includes any. A green smoothie is a perfect way to start the day with some vegetables. We usually prepare all the ingredients the night before and refrigerate them (unblended) in a blender jar so that a quick blend is all that is required in the morning.

INGREDIENTS:

1 pear (in fall or winter) or peach (in summer), seeds or pit removed but skin left on

1 banana
2 cups leafy greens (spinach, lacinato kale with stems removed, or beet greens)
1 cup of plain, unsweetened kefir or yogurt
1 teaspoon vanilla extract
½–1 cup water
Ice cubes (optional; add just before blending)

INSTRUCTIONS:
Add the fruit, greens, kefir or yogurt, vanilla, and water together in a blender. Blend until smooth, adding additional water and/or ice to get the consistency desired.

Bacteria-Boosting Granola

SERVES 8
(6 grams of fiber per serving, not including added fruit)

Store-bought granola usually has an overabundance of added sugar, which is unfortunate because granola has great potential for supporting microbiota health. This recipe keeps all the great dietary fiber but eliminates much of the added sugar. To keep it interesting throughout the year, we often add seasonal fruit when we eat the granola.

INGREDIENTS:
4 cups mixed rolled cereal grain (or 1 cup each of flakes from oats, barley, rye, and quinoa; substitute as desired; Bob's Red Mill brand has a five-grain rolled cereal that works well)
1 cup unsweetened dried flake coconut
1 cup chopped almonds
½ cup pepitas (pumpkin seeds)
½ cup pumpkin puree

3 tablespoons olive oil
½ cup water
2 tablespoons maple syrup
1 teaspoon ground cinnamon
1 teaspoon vanilla extract
½ cup raisins

INSTRUCTIONS:
Preheat the oven to 350°F. In a large bowl, mix together the rolled cereal, coconut, almonds, and pepitas. In a small bowl, whisk together the pumpkin puree, olive oil, water, maple syrup, cinnamon, and vanilla. Pour the wet mixture over the cereal mixture and stir to coat. Spread the mixture onto a large baking sheet and bake for 40 minutes or until golden brown, stirring halfway through. Add the raisins to the cooked granola. When cooled, store the granola in a covered container in the refrigerator. Granola keeps well for at least a month. Serve about ¾ cup over yogurt or kefir with seasonal fresh or thawed frozen fruit.

Muesli for Microbes

SERVES 4
(7.5 grams of fiber per serving)

Dr. Maximilian Oskar Bircher-Benner was a Swiss physician who ran a sanatorium in Zurich in the late 1800s. He believed that a diet rich in fruits, vegetables, and nuts could be used to help heal his patients. He formulated the cereal muesli, which is still referred to as Bircher muesli in much of Europe. The original Bircher muesli breakfast contained an entire grated apple with only a couple tablespoons of cereal soaked in apple juice. Here we have adapted his recipe, but we have tried to maintain the higher fruit-to-grain ratio Dr. Bircher originally intended.

INGREDIENTS:

4 apples, unpeeled and chopped
2 cups plain, unsweetened kefir
½ cup mixed rolled cereal
¼ cup chopped hazelnuts
2 tablespoons flaxseed meal
2 tablespoons lemon juice
¼ teaspoon nutmeg
¼ teaspoon sea salt
Raw honey

INSTRUCTIONS:

Chop the apples by hand or in a food processor. Combine the kefir, cereal, hazelnuts, flaxseed meal, lemon juice, nutmeg, salt, and chopped apples in a large bowl. Store the bowl in the refrigerator overnight. Serve with a small drizzle of raw honey. Muesli can be stored in the refrigerator for a few days.

Tarahumara Pancakes

SERVES 4

(9 grams of fiber per serving, not including added fruit)

Weekend pancakes are as much an institution in American families as the Saturday morning cartoons that often accompany them. Most kids love pancakes, and who can blame them? Highly refined flour and copious amounts of maple syrup or powdered sugar (or both!) make this breakfast staple a close cousin to the cupcake. We've taken inspiration for this recipe from the

Tarahumara of Mexico. These Native Americans are known for their incredible stamina that enables them to run for hours on end, their good health, and their high consumption of dietary fiber. They make something called pinole that contains stone-ground cornmeal, chia seeds, and some spices that can be made into a beverage or baked into a little cake. Coarsely ground cornmeal helps to deliver more of the carbohydrates to the microbiota. This microbiota-mindful pancake will leave you feeling full and with enough energy to propel you through a fun and productive Saturday.

INGREDIENTS:

3/4 cup medium ground cornmeal
1 cup boiling water
1 cup whole grain wheat flour
1/4 cup chia seeds
1 tablespoon ground cinnamon
1 1/2 teaspoons baking powder
1/2 teaspoon baking soda
1/2 teaspoon sea salt
1 1/2 cups buttermilk
1 teaspoon vanilla extract
4 tablespoons olive oil
Plain unsweetened yogurt
Berries
Maple syrup

INSTRUCTIONS:

Add the boiling water to the cornmeal in a large bowl. In a medium bowl, stir together the flour, chia seeds, cinnamon, baking powder, baking soda, and salt. Add the buttermilk, vanilla, and olive oil to the large bowl containing the softened cornmeal. Stir the dry ingredients into the wet ingredients until evenly combined. Bring an oiled medium skillet to medium heat, scoop

1/4-cup spoonfuls of the batter into the pan, and cook the pancakes on both sides until browned. Serve with yogurt, berries, and a drizzle of maple syrup.

Symbiotic Scramble

SERVES 4

(13 grams of fiber per serving)

This healthier take on the breakfast burrito has plenty of beans and vegetables to feed the microbiota and omits the cheese in favor of probiotic-containing Greek yogurt. Try to find corn tortillas without added extra ingredients; ideally, the ingredient list should contain only corn flour, water, lime, and salt. If you are feeling really adventurous you can try to make tortillas at home—but remember that once you do this, you'll never again be satisfied with the store-bought alternative. All you need is a bag of masa harina (corn flour) and a tortilla press or rolling pin. A freshly made warm tortilla is a true treat. Serve with an Aztec Hot Chocolate for a leisurely Sunday breakfast.

INGREDIENTS:

1 onion, chopped
6 large eggs
Sea salt and black pepper, to taste
2 cups black beans
8 corn tortillas
1 avocado, cubed
Salsa and Greek yogurt to serve

INSTRUCTIONS:

Warm an oiled medium skillet on medium heat. Cook the onions until soft, about 4 minutes. Beat the eggs together in a medium bowl with salt and pepper. Add the eggs to the pan with the onions and cook while stirring. When the eggs are almost set, add the beans and stir until the eggs are cooked through. Wrap the tortillas in a damp cloth and warm them in the microwave until they become pliable, approximately 30 to 60 seconds on high followed by a few minutes' rest in the cloth. Serve the eggs with avocado cubes, salsa, yogurt, and two warm corn tortillas.

Aztec Hot Chocolate

SERVES 2

(4 grams of fiber per serving)

The ancient Aztecs drank a chocolate beverage made with water, cocoa, and chili peppers called *xocolatl*, which translates as "bitter water." This drink was a far cry from the overly sweetened whipped-cream-topped confection that we usually think of as hot chocolate. This microbiota-friendly breakfast drink is not bitter and is great for kids and, with an added shot of espresso, for adults.

INGREDIENTS:

2 cups milk or unsweetened almond milk
¼ cup unsweetened organic cocoa powder
Pinch sea salt
1 teaspoon molasses
1 teaspoon ground cinnamon
¼ teaspoon vanilla extract
Pinch cayenne (optional)
Grated orange peel (optional)

INSTRUCTIONS:

Add the milk, cocoa powder, salt, molasses, cinnamon, vanilla, and cayenne (if using) to a small saucepan. Whisk over medium heat until the cocoa powder is incorporated and the mixture is hot, about 5 minutes. Be sure not to let the milk scald. Alternatively, add the ingredients to a large microwave-safe bowl and microwave until hot, pausing frequently to whisk. Top with grated orange peel.

School LUNCHES EVEN THE MICROBIOTA CAN LOVE

Packing a healthy school lunch that our children will eat involves a delicate balancing act of marrying highly nutritious ingredients with flavors, textures, and colors that appeal to kids. Many schools are improving the selection of healthy food they offer, but often kids don't make the healthiest choices. And who can blame them—a cheesy burger on a cake-like bun can be a lot more appealing to a child than a salad. By packing our kids' lunches we make sure they see firsthand what a healthy meal looks like, and we always try to use fresh (and thereby tasty) ingredients. We have only included a couple of school lunch suggestions since we prefer to pack smaller portions of the lunches we prepare for ourselves. Developing their palate for "grown-up" food helps kids eat more nutritiously, in general.

PB&J 2.0

(6–8 grams of fiber per sandwich, depending on fruit used)

Here is a microbiota-reframed PB&J recipe that is easy to pack in a lunch box. Check the ingredient list on nut butters and avoid those with added sugar, palm oils, or too much salt. Finding one that fulfills these criteria can be challenging. Nut butter is easy to make, and making it yourself ensures that it is fresh and does not contain unwanted additives. We do this routinely and it allows us to try a variety of nut combinations. This mixture is a favorite of ours.

Almond/Walnut Butter

INGREDIENTS:

1 cup unsalted almonds (raw or roasted)

1 cup walnuts

1 tablespoon extra-virgin olive oil

INSTRUCTIONS:

Add the almonds, walnuts, and olive oil to a food processor or high-powered blender and blend for a few minutes until smooth. Store in a mason jar in the refrigerator. For sandwiches, use whole grain bread with homemade nut butter and slices of seasonal fruit instead of jam. The whole fruit provides the sweetness and the necessary fiber without the added sugar of jam. The choice of fruit can be tailored to what is seasonally available and will make the freshest tasting sandwich. Thinly sliced apples or pears with the peel left on are great in fall and winter. In spring, sliced strawberries, peaches, or nectarines can be used. Bananas are a good fall-back option and are available all year. If your children are hooked on the sweetness of jelly, you can add a drizzle of honey over the fruit to start, and then slowly eliminate this step over time—done gradually enough, your kids probably won't even notice.

SERVING SUGGESTIONS:

Provide carrot sticks, red pepper slices, cucumber slices, or celery sticks and plain (unsweetened) Greek yogurt for dipping.

We stay away from the low-fat or nonfat kinds of yogurt. Whole milk yogurt tastes much better and there is evidence that dairy fat, especially from organic milk, has health benefits and keeps kids leaner in the long run.

As a treat, skip the cookie and include a small square of dark chocolate (70 percent or higher cocoa is best). Dark chocolate has many healthful compounds as well as almost 2 grams of dietary fiber per ounce.

Big "MAC" Quesadillas

SERVES 2

(9 grams of fiber per quesadilla)

Most kids love quesadillas made with pale white-flour tortillas and lots of melty cheese, which provides little for the microbiota. However, with just a few tweaks quesadillas can become great microbiota food. Using stone-ground corn tortillas provides more fiber than the refined wheat flour variety. Adding black beans really elevates the simple quesadilla into something healthy and delicious. Many people presoak their beans overnight before cooking them, but we find this step unnecessary. To cook black beans at home, sort through a 1-pound bag spread on a baking sheet to pick out any stones or dirt clumps, pour the beans into a large pot, and cover them with about 3 inches of water and a pinch of salt. Bring the water to a boil, then lower the heat and simmer the beans, covered, for about 2 hours or until they reach the desired tenderness. Add water, if needed, as they simmer. We store cooked, drained beans in mason jars in the refrigerator or freezer.

INGREDIENTS:

1 cup black beans, drained

¼ cup cheddar or Monterey Jack cheese, shredded

1 teaspoon cumin

2 stone-ground corn tortillas†

Cilantro, chopped

INSTRUCTIONS:

Combine the beans, cheese, and cumin in a small bowl. Cover half of one tortilla with half of the bean mixture and fold it in half. Warm an oiled medium skillet on medium heat and cook the quesadilla on both sides until the cheese is melted, then remove it from the pan and insert the cilantro. Repeat with the second tortilla.

† Store-bought corn tortillas can be made more pliable and fold more easily by prewarming on the skillet or wrapping them in a damp cloth and warming them in the microwave.

SERVING SUGGESTION:

Serve with cubes of avocado and cherry tomatoes on the side. Include a seasonal piece of fruit for dessert.

Work LUNCH

Many workplaces—including some medical schools we've worked at!—don't offer good lunch choices. It's often better for your health and your wallet to bring a packed lunch, which we do almost daily. We like to pack our lunch the night before, just as we do for our kids, to make the morning a little easier. Often our lunch is just leftovers from the previous night's dinner. But if there aren't any leftovers, we put together a quick salad filled with fresh vegetables, grains, beans, nuts, and seeds. These dishes are also easy to put in a child's lunch box.

Chickpea Greek Salad

SERVES 2

(16 grams of fiber per serving)

To cook chickpeas at home, follow the instructions in the previous recipe for preparing black beans.

INGREDIENTS:

2 cups or one 15-ounce can cooked chickpeas
1 cucumber, sliced, unpeeled
1 cup cherry tomatoes
½ purple onion, sliced
1 green bell pepper, chopped
½ cup fresh parsley, chopped
¼ cup pitted Kalamata olives, chopped
¼ cup feta cheese crumbles
Lemon juice
Extra-virgin olive oil
Freshly ground black pepper

INSTRUCTIONS:

Toss the chickpeas, cucumber, tomatoes, onion, pepper, parsley, olives, and feta together. Dress with fresh-squeezed lemon juice and extra-virgin olive oil. Top with freshly ground black pepper.

Soba Noodle Salad with Probiotic Peanut Miso Sauce

SERVES 4

(11 grams of fiber per serving)

Soba noodles are a type of Japanese noodle made from buckwheat, which, despite its name, is not a variety of wheat.

Noodles made from 100 percent whole grain buckwheat will have the most fiber. The sauce contains miso, which is a paste made from fermented soybeans with added barley or rice. Miso is fermented using a starter that contains a species of fungus called *Aspergillus oryzae*. If you can find unpasteurized miso you will benefit not only from the fermentation products it contains but also from its living microbes (pasteurization kills these microbes). This dish comes together easily and is perfect for a child's school lunch since it is served cold or at room temperature.

INGREDIENTS:

One 9-ounce package of 100 percent whole grain soba noodles
4 cups carrots, grated
4 cups shelled and cooked edamame
2 cups radishes, sliced
1 cup scallions, sliced
¼ cup sesame seeds

PEANUT MISO SAUCE:

¼ cup sesame oil
¼ cup water
¼ cup soy sauce
4 tablespoons peanut butter
2 tablespoons unpasteurized miso paste (white or yellow)
1 tablespoon freshly grated ginger†
1 tablespoon sugar
1 lime, juiced

INSTRUCTIONS:

For the sauce: Combine the ingredients in a blender and blend until smooth.

For the salad: Boil the soba noodles in a medium saucepan until al dente, about 3 to 4 minutes. Drain noodles and cool them under running cold water. In a large bowl combine the noodles with the carrots, edamame, radishes, and scallions, and toss with peanut miso sauce. Serve cold or at room temperature with the sesame seeds sprinkled on top.

† We keep a root of ginger in the freezer. The frozen root keeps for months and is easily grated with a Microplane grater.

Totally MAC-Filled Tabbouleh

SERVES 4

(7 grams of fiber per serving)

Bulgur wheat is one of the higher-fiber-containing grains. It is made from wheat that has been hulled but still maintains most of its bran. It is usually sold parboiled and dried, which makes it a quick-cooking whole grain. This grain has been a staple in the Mediterranean diet, which remains one of the world's healthiest with its emphasis on low saturated fat, high fiber, and fermented foods. Tabbouleh is traditionally served as a salad or side dish, but the addition of chickpeas transforms it into a satisfying complete meal, one that would be a refreshing dinner on a hot day. Nuts or seeds can be added for additional fiber and crunch.

INGREDIENTS:

1 cup bulgur wheat
1 cup boiling water
1 cup cooked chickpeas
1 cup fresh parsley, minced
1 cucumber, chopped
2 celery stalks, chopped

2 medium tomatoes, chopped
¼ cup sliced green onions
Plain, unsweetened yogurt

DRESSING:
1 lemon, juiced
3 tablespoons extra-virgin olive oil
1 clove garlic, crushed
1 teaspoon ground allspice
Freshly ground black pepper

INSTRUCTIONS:
Pour the boiling water over the bulgur in a medium bowl and let stand until the water is absorbed, approximately 15 minutes. In a large bowl combine the chickpeas, parsley, cucumber, celery, tomatoes, and green onions. Add the cooked bulgur and stir to combine. In a small bowl, whisk together the lemon juice, olive oil, garlic, allspice, and black pepper and incorporate the mixture into the salad. Serve immediately or refrigerate briefly to cool. Garnish with a dollop of plain yogurt.

Snacks

We have mixed feelings about snacking for kids. On the one hand we understand that children have small stomachs and often find it hard to make it to the next meal without a little something in between. However, we have also seen firsthand how snacking can be used by children to avoid the healthier fare that is available at mealtimes. So here is how we handle after-school snacking to ensure our kids will be hungry at mealtimes but make it through the afternoon without a major meltdown.

First, they need to finish whatever lunch they haven't eaten at school. If you use reusable packaging[†] for school lunches it's easier to see what they've eaten and what's remaining. As with most kids, if there are leftovers, it's usually the vegetables. Often just finishing their lunch is enough to get them through until dinner. If they are still hungry after finishing their lunch food, then we provide a healthy snack that isn't too filling so that they are hungry when dinner is served. We have found that minimizing snack food is critical to getting our kids to eat a healthy dinner without a fight. Kids are not going to be as excited about dinner if their tummies are full from too much snack food. Hunger, as they say, is the best seasoning.

For adults, bringing homemade snacks to work is also a good way to ward off desperate late-afternoon trips to the vending machine—usually not a good source of healthy food. Often a piece of fruit or a handful of nuts is the simplest solution, but here are a few other ideas for snacks that keep the microbiota in mind.

[†] Klean Kanteen and LunchBots make great stainless steel lunch containers.

Hunter-Gatherer Tuber Snack

SERVES 8

(8 grams of fiber per serving)

Homemade hummus is wonderful, especially if it's made from home-cooked chickpeas, but there are also good store-bought options that don't contain a lot of unrecognizable ingredients. Hummus is a healthy, microbiota-nourishing dip that most children enjoy. We often use hummus as a dip for a variety of vegetables, including jicama. Jicama is a tuber that contains a good amount of fiber (although not nearly as much as the tubers the Hadza eat). By pairing this tuber with hummus you boost the snack's fiber content to more closely resemble the wild tubers that hunter-gatherers eat. If you tell your kids that tubers helped to sustain hunter-gatherers before farming was developed, you may find them happily dipping and crunching this Hadza-inspired snack.

INGREDIENTS:

2 cups or one 15-ounce can of cooked chickpeas
1 lemon, juiced†
¼ cup tahini
2–3 tablespoons water
1 clove garlic, crushed
Sea salt to taste
Extra-virgin olive oil
Pinch paprika
Jicama, peeled and cut into sticks

INSTRUCTIONS:

Add the chickpeas, lemon juice, tahini, water, garlic, and salt to a food processor or blender and blend until smooth. Add water as needed to achieve the desired consistency. Serve the hummus in a bowl, topped with a drizzle of olive oil and a pinch of paprika alongside a plate of jicama sticks.

† Add a bit of grated lemon zest for more intense lemon flavor.

Probiotic Pick-Me-Up

Most supermarket yogurt is loaded with added sugar, making what should be a healthy snack more of a pudding-like dessert. Many times the plain varieties are only available in flavorless low-fat or nonfat versions. Plain whole milk yogurt is deliciously tangy and a small amount can be very satisfying. It is worth seeking out organic plain whole milk yogurt or, if you are interested in becoming a budding microbiologist, fermenting your own. Here is our method for making homemade yogurt—a great project for teaching your kids about the magic of microbes. If you get adventurous you can experiment with yogurt cultures from all over the world; they're available at culturesforhealth.com.

Plain Yogurt

INGREDIENTS:

1 quart organic whole milk
Approximately 1/4 cup yogurt, or contents of a yogurt culture packet

INSTRUCTIONS:

Heat the milk in a medium saucepan to 180°F while occasionally stirring. Carefully monitor the temperature of the milk so that it doesn't boil over or scald. Once it has reached temperature, remove the pan from the heat and allow the milk to cool to 115°F. Add the yogurt or yogurt culture to the warm milk, whisk, and transfer to a 1-quart mason jar and close the lid tightly. Place the jar in a yogurt maker or an insulated cooler with enough warm water (105–115°F) so that the jar is sitting in a couple inches of water. Leave the yogurt overnight to ferment, then put it in the refrigerator the next morning to firm up.

Crunchy Yogurt Parfait

SERVES 1

(7 grams of fiber per serving)

INGREDIENTS:

½ cup plain, unsweetened yogurt

½ cup mixed berries, fresh or defrosted frozen

¼ cup chopped hazelnuts

INSTRUCTIONS:

Layer the fruit and nuts on top of the yogurt. Stirring it all together is optional.

Japanese Popcorn

SERVES 4

(2.5 grams of fiber per serving)

Even if you don't have a microbiome with seaweed-degrading porphyranases there are still many reasons to eat sea vegetables. They are loaded with an assortment of minerals and have a complex seafood flavor. Sprinkle nori on fiber-loaded popcorn (which is a whole grain) to make a healthy treat for you and your microbiota.

INGREDIENTS:

2 tablespoons sesame oil, divided

⅓ cup popcorn kernels

2 sheets nori, crushed

½ teaspoon sea salt

1 teaspoon wasabi powder or cayenne, optional

INSTRUCTIONS:

In a large pot heat 1 tablespoon of oil on high heat, add the popcorn kernels, and cover. Once the kernels begin to pop, vigorously shake the pan. When the popping subsides, remove the pot from the heat immediately to avoid burning the popcorn. Transfer the popcorn to a large rimmed baking sheet. Drizzle the remaining 1 tablespoon of sesame oil over the popcorn and sprinkle it with the crushed nori and optional wasabi powder or cayenne for some heat. Toss well and serve.

Cashews for Your Commensals

MAKES 4 CUPS

(3 grams of fiber per cup)

Turmeric is medicinal in its anti-inflammatory effects. We try to incorporate this spice into dishes whenever we can. Paired with cashews, this Middle Eastern–style snack is both tasty and filled with microbiota accessible carbohydrates.

INGREDIENTS:

1 tablespoon olive oil
4 cups raw cashews
1 tablespoon turmeric
1 teaspoon sea salt

INSTRUCTIONS:

Heat the oil in a medium skillet over medium/high heat. Add the cashews and sprinkle with sea salt. Roast the cashews while stirring frequently, about 5 minutes, then remove from the heat and toss in the turmeric. Let cool and serve. These may be stored in a jar with a tight-fitting lid.

Fermented-Filling Dates

SERVES 4

(3.5 grams of fiber per serving)

Medjools are often called the king of dates. They can be expensive but they have such a satisfying taste that just a couple can keep your appetite appeased until dinner. Stuffed with probiotic-containing cultured cream cheese, these snacks will also provide some visiting microbes to keep your resident microbiota happy.

INGREDIENTS:

8 Medjool dates
3 tablespoons cultured cream cheese
8 walnut halves
Ground cinnamon, optional

INSTRUCTIONS:

Slit open one side of the date to remove the pit. Stuff a generous teaspoon of cream cheese and a walnut half into each date. Serve with an optional sprinkle of cinnamon.

Dinner

Dinner can be a struggle for families. By the end of the day everyone is tired and the last thing parents want is to cook a complicated meal and struggle with their children over eating it. Here we try to provide recipes that can be made relatively quickly and easily on a weeknight, furnish enough nourishment for the microbiota, and also be something kids will enjoy. Keep in mind that it's worth introducing these dishes over and over again, even if your children don't seem to like them at first. Over time they will develop a taste for healthier food—a wonderful gift for them throughout their lives.

Mutualist Mediterranean Soup

SERVES 6

(8 grams of fiber per serving)

During the colder months we make soups at least a couple of times per week. They warm you up and have a reputation for keeping colds at bay. Soup is a great way to begin reintroducing legumes into your diet if they are not there already. You can adjust this recipe to slowly increase the amount of beans you add to gradually boost the fermentation in your gut.

INGREDIENTS:

2 tablespoons extra-virgin olive oil
1 purple onion, diced†
1 fennel bulb, chopped†
4 cups chopped lacinato kale, leaves and stems separated†
4 cloves garlic, crushed
4 cups low-sodium vegetable broth
2 cups water
1 bay leaf
2 cups or one 15-ounce can cooked cannellini beans
2 cups diced tomatoes
1 cup sliced carrots
Sea salt and pepper, to taste

INSTRUCTIONS:

Add the olive oil to a Dutch oven or large stockpot and heat over medium/high heat. Add the onion, fennel, and chopped kale stems and cook until softened, about 6 minutes. Add the garlic and cook for an additional minute. Add the broth, water, bay leaf, beans, tomatoes, and carrots and simmer covered for about 15 minutes. Season with salt and pepper to taste. Serve with grated Parmesan cheese, freshly ground black pepper, and a thick slice of whole wheat sourdough bread with extra-virgin olive oil.

† Using a food processor can help speed up the prep work.

Sesame Seed-Crusted Salmon with Green Beans and Orange Miso Sauce

SERVES 4

(9 grams of fiber per serving; 13 grams if served with 1 cup of brown rice)

Seeds are a wonderful source of dietary fiber, healthy oils, protein, and various micronutrients. We keep a variety of seeds in our pantry to sprinkle on salads, cooked vegetables, hot cereal, and even yogurt. This is an easy recipe to pull together on a weeknight but is also sophisticated enough for a dinner party.

INGREDIENTS:

Four 4-ounce fillets of wild salmon
½ cup sesame seeds
2 tablespoons olive oil, divided
2 pounds green beans, washed and ends trimmed
1 cup slivered almonds
Sea salt and pepper

ORANGE MISO SAUCE:

1 cup orange juice

2 tablespoons unpasteurized miso paste (white or yellow)

1 tablespoon sesame oil

1 tablespoon grated ginger

1 tablespoon grated orange peel

INSTRUCTIONS:

Spread the sesame seeds on a large plate and press the salmon fillets onto the seeds to coat them. Heat 1 tablespoon of olive oil over medium heat in a large skillet. Add the salmon and cook approximately 4 minutes on each side until cooked through completely. Remove the fillets from the skillet and cover with foil to keep them warm.

Coat the skillet with the remaining olive oil and add the green beans and almonds. Sauté on medium heat until the green beans are cooked but still crisp and the almonds are slightly toasted, about 5 minutes. Season with salt and pepper.

For the sauce: blend the orange juice, miso paste, sesame oil, grated ginger, and orange peel. Serve the salmon and green beans with brown rice and the sauce.

Fiber-Filled Flatbread Pizza

SERVES 4

(9 grams of fiber per serving)

There is an almost universal love of pizza among children, which makes this dish a perfect way to introduce them to more vegetables. You can pick up a whole wheat flatbread at the grocery store (we like to use whole wheat naan). Slather on a layer of parsley almond pesto and top as you like. We have provided some ideas for high-fiber toppings. Bake it on a baking sheet or a pizza stone for an extra-crispy crust.

Parsley Almond Pesto

INGREDIENTS:

2 cups fresh parsley
2 cloves garlic
½ cup unsalted almonds
2–3 tablespoons extra-virgin olive oil
1 tablespoon lemon juice
Sea salt to taste

INSTRUCTIONS:

Add parsley, garlic, almonds, olive oil, lemon juice, and salt to a food processor and blend until smooth.

Pizza

INGREDIENTS:

4 whole wheat flatbreads†
Parsley Almond Pesto (above recipe)
1 cup sliced sun-dried tomatoes
1 cup chopped artichoke hearts
1 medium onion, sliced
¼ cup Kalamata olives
2 tablespoons capers
Parmesan cheese, grated, to taste

INSTRUCTIONS:

Preheat the oven to 550°F. Spread a thin layer of pesto onto the flatbread, then garnish it with the tomatoes, artichoke hearts, onion, olives, capers, and grated Parmesan cheese. Place the flatbread on a jelly roll pan, cookie sheet, or pizza stone on the bottom shelf of the oven and bake it until it is heated through, about 8 minutes. Remove the pizza from the oven and allow it to cool slightly, then cut and serve.

† If you are feeling adventurous, several recipes are available online, including 100 percent whole wheat pizza dough from www.foodnetwork.com and one in the Recipes for Health section of www.nytimes.com. Increase cooking time to 10 minutes if the crust is not pre-baked. Remember to use whole grain flour.

Microbiota-Reframed Risotto

SERVES 4

(19 grams of fiber per serving)

Risotto is traditionally made using the starchy white Arborio rice that has a high glycemic load and only 1 gram of fiber per half cup serving. Barley, on the other hand, has a very low glycemic load and almost 5 grams of fiber per serving, not to mention that it's loaded with molecules thought to help lower cholesterol. This risotto also has lots of high-fiber additions such as artichoke hearts and oyster mushrooms.

INGREDIENTS:

1 cup barley
1 tablespoon extra-virgin olive oil
1 large onion, chopped
1 clove garlic
1 cup chopped oyster mushrooms
1 cup chopped artichoke hearts

1 cup diced tomatoes
1 cup chicken or vegetable stock
¼ cup grated Parmesan cheese
Fresh oregano
Sea salt and pepper, to taste

INSTRUCTIONS:

Bring 2½ cups of water to boil in a medium saucepan. Add the barley and simmer for 10 minutes. Meanwhile, in a large sauté pan heat the olive oil over medium heat. Add the chopped onions and cook until it begins to turn brown, about 5 minutes. Add the garlic and cook for an additional minute. Add the mushrooms, artichoke hearts, and tomatoes and cook until the mushrooms are soft. Add the stock and reduce for 5 minutes. Stir in the cooked barley, cheese, and oregano and add salt and pepper to taste.

Indian Dal

SERVES 4

(10 grams of fiber per serving)

Cultures that have a long tradition of consuming legumes often couple them with spices that supposedly reduce their gas-producing quality. Dal is a lentil dish that is consumed throughout South Asia. Lentils are a great way to incorporate legumes into a weeknight dinner since they cook so quickly compared to dried beans. Serve them on top of a wild rice blend with a Mango Kefir Lassi (following recipe) and your microbiota and palate will thank you.

INGREDIENTS:

3 tablespoons extra-virgin olive oil
1 teaspoon mustard seeds
1 teaspoon grated fresh ginger

2 cloves garlic, minced
1 onion, chopped
5 medium carrots, chopped
5 celery stalks, chopped
2 teaspoons ground coriander
1 teaspoon turmeric
1 teaspoon ground cumin
1 teaspoon cayenne pepper
½ teaspoon ground cinnamon
½ teaspoon ground cloves
1½ cups red lentils, rinsed
4 cups vegetable stock or water
One 18-ounce can diced tomatoes
1 teaspoon sea salt
1 lime, juiced
½ cup fresh cilantro, chopped

INSTRUCTIONS:

Heat the olive oil in a large sauté pan or Dutch oven over high heat. Add the mustard seeds and fry until they begin to pop, about 1 minute. Reduce the heat to medium and add the ginger, garlic, onion, carrots, and celery. Cook until the vegetables just begin to soften, about 5 minutes. Add the coriander, turmeric, cumin, cayenne, cinnamon, and cloves and stir to combine. Add the rinsed lentils, stock or water, diced tomatoes, and salt. Stir, and bring to a boil. Reduce heat and simmer, covered, stirring occasionally, for about 15 to 20 minutes, or until the lentils are done and the soup has thickened. Add the lime juice, cilantro, and salt to taste before serving.

Mango Kefir Lassi

SERVES 2

(3 grams of fiber per serving)

INGREDIENTS:

2½ cups plain, unsweetened kefir
2 cups frozen mango chunks
1 teaspoon ground cardamom
1 teaspoon chopped fresh mint
1 tablespoon honey
Water and/or ice for desired consistency

INSTRUCTIONS:

Combine the kefir, mango, cardamom, mint, honey, and water in a blender and blend until smooth. Add more water and/or ice to achieve the desired consistency.

Desserts

Most nights we have some sort of dessert. It helps motivate our kids to finish their "growing" food and it's a nice way to end the day. We keep these desserts very simple—a small bowl of fresh berries, pear wedges, or a simple piece of dark chocolate. For a special occasion we will make a more involved dessert.

Microbe-Friendly Oatmeal Cookies

MAKES 2 DOZEN COOKIES

(1 gram of fiber per cookie)

These cookies are a version of the standard oatmeal chocolate chip recipe but modified to better feed the microbiota. We replace the chocolate chips with cacao nibs, which are pieces of cacao beans that have been fermented and roasted. (Cacao nibs are not a source of live active cultures because the organisms don't survive the roasting process.) Their taste is nutty and is reminiscent of coffee beans. There is very little flour in these cookies; they're mostly high-fiber rolled oats.

INGREDIENTS:

1 tablespoon whole grain wheat flour
1 teaspoon baking powder
$\frac{1}{4}$ teaspoon sea salt
1 teaspoon ground cinnamon
2 tablespoons cacao nibs, crushed
$\frac{1}{4}$ cup ($\frac{1}{2}$ stick) unsalted, cultured butter
$1\frac{1}{2}$ cups rolled oats (not instant)
$\frac{1}{4}$ cup olive oil
$\frac{1}{3}$ cup sugar
1 large egg

INSTRUCTIONS:

Preheat the oven to 350°F. Line a baking sheet with parchment paper or a Silpat. In a small bowl, combine the flour, baking powder, salt, cinnamon, and cacao nibs. In a separate bowl, melt the butter in the microwave, then stir the oats and olive oil into the melted butter. In a separate large bowl, whisk the sugar with the egg until creamy. Add the flour mixture and the oats mixture to the egg mixture and stir until combined. Drop approximately 1 tablespoon of dough onto baking sheets for each cookie. Bake the cookies 8 to 10 minutes, until golden brown.

Brownies for Your Bacteria

MAKES 16 BROWNIES

(2 grams of fiber per brownie)

At some point chocolate got a bad rap, which it rightly deserves when consumed in its ultrasweetened milk chocolate candy bar form. But more and more studies are showing that dark chocolate that contains at least 70 percent cocoa can be a healthy treat due to the presence of flavonoids. Chocolate also has another magic ingredient, fiber. A 1.5-ounce serving of dark chocolate contains about 3 grams of dietary fiber. In this recipe we combine chocolate and another rising star in the healthy food category, nuts, for a brownie treat that kids, adults, and their microbes will love.

INGREDIENTS:

5 tablespoons unsalted, cultured butter
6 ounces dark chocolate (70 percent cocoa)
1 cup almond meal
⅓ cup sugar
1 tablespoon cacao nibs
2 large eggs
1 teaspoon vanilla extract
1 teaspoon ground cinnamon
1 teaspoon sea salt
1 tablespoon orange zest

INSTRUCTIONS:

Preheat the oven to 350°F. Melt the butter and chocolate in the microwave, stirring occasionally to make sure the chocolate doesn't burn. Add the almond meal, sugar, cacao nibs, eggs, vanilla, cinnamon, salt, and orange zest and whisk until all ingredients are incorporated. Pour into an 8" x 8" oiled baking pan. Bake for 30 minutes or until a toothpick inserted in the center comes out clean.

Burkina Faso Skillet Cake

SERVES 6

(4 grams of fiber per serving)

This recipe was inspired by the study comparing the microbiota of children from Burkina Faso in West Africa to that of Italian children. Many people in Burkina Faso eat a diet that is much higher in fiber than the average Westerner's and includes millet, sorghum, legumes, nuts, fruits, and veggies. In this recipe we use millet to make the base for a skillet cake.

INGREDIENTS:

½ cup millet
1 cup water
2 tablespoons (¼ stick) unsalted, cultured butter
4 cups berries, fresh or frozen
¾ cup whole grain wheat flour
1 teaspoon baking powder
¼ teaspoon baking soda
Pinch sea salt
¾ cup plain, unsweetened kefir or cultured buttermilk
1 large egg
¼ cup molasses
6 tablespoons extra-virgin olive oil
1 teaspoon vanilla extract
¾ cup chopped unsalted peanuts
Plain, unsweetened yogurt for garnish

INSTRUCTIONS:

Preheat the oven to 400°F. In a small saucepan add 1 cup of water to the millet and bring it to a boil, then lower the heat and simmer for 15 minutes, covered. Heat a 10-inch skillet with an ovenproof handle over medium heat and add the butter. When it begins to brown, add the fruit. Cook, stirring, until the fruit is soft, about 5 minutes depending on the type of fruit used. In a medium bowl stir together the flour, baking powder, baking soda, and salt. To the dry mixture add the cooked millet, kefir or buttermilk, egg, molasses, olive oil, and vanilla and whisk until a homogenous batter is formed. Pour the batter over the cooked fruit and bake in the oven for 25 to 30 minutes or until a toothpick inserted into the cake comes out clean. Allow the skillet to cool for about 10 minutes, then invert the cake over a plate. Sprinkle peanuts over the top and serve with a dollop of yogurt.

Middle Eastern Oatmeal Pudding

SERVES 4

(9 grams of fiber)

Steel-cut oats are the less processed sibling of the more familiar rolled oats. They are made from whole oat groats that are only cut, whereas rolled oats have been steamed and rolled. Quick-cooking oats are rolled even thinner than rolled oats and instant oats are thinner still. While the nutritional facts state that the fiber content for all these manifestations of oats is the same, the uncrushed steel-cut oats deliver more of the carbohydrate content to the microbiota. Also steel-cut oats have a more toothy texture, which in our opinion makes them a little more interesting to eat. This pudding also makes a great breakfast.

INGREDIENTS:

4 cups water
1¼ cups steel-cut oats
Pinch sea salt
1 cup golden raisins
1 cup pistachios, chopped
1 tablespoon unsalted, cultured butter
½ teaspoon ground cardamom
1 tablespoon honey
Plain, unsweetened kefir or yogurt

INSTRUCTIONS:

Bring 4 cups of water to a boil in a medium saucepan. Add the oats and a pinch of salt and cook for about 20 minutes over medium heat, stirring occasionally. Add the raisins, pistachios, butter, cardamom, and honey during the last 5 minutes. Serve with kefir or yogurt as desired.

ACKNOWLEDGMENTS

First we'd like to thank our editor, Virginia "Ginny" Smith; Ann Godoff; and the team at Penguin Press for their vital role in helping us realize our shared vision in this endeavor. Ginny was with us every step and word of the way, and aided us immensely in shaping this book. Dr. Andrew Weil was the catalyst for this project. He saw the need for this new research to be made accessible to all and we thank him for his encouragement and guidance. Richard Pine, our agent, was critical in helping us navigate the unfamiliar terrain of the publishing world. We are grateful for his sage advice throughout this process.

We thank our wonderful colleagues at Stanford and around the world for many enlightening discussions and ideas. Our mentors and teachers along the way are too numerous to mention by name, but we are grateful for all that each of them has given to us. The research covered in this book is the collective effort of many scientists working to decode the mysteries of the human microbiota. Their creativity, intelligence, and tenacity inspired not only this book but also our continued work in this ever-expanding field of medical research. We are particularly indebted to the scientists mentioned in the preceding pages for taking the time to speak to us about their research and the field in general. Numerous colleagues read through portions of this book and provided valuable comments and fact-checking, including Kristen Earle, Jon Lynch, Angela Marcobal, Katharine Ng, Sam Smits, Liz Stanley, and Weston Whitaker.

We are fortunate to work in a field with so many brilliant, generous, and collaborative people. Our mentor Dr. Jeffrey Gordon deserves special praise in this regard. He ignited our passion for the microbiota years ago and his continued work in this area serves as a source of constant awe and admiration. We also humbly thank the past and present members of our laboratory at Stanford. Their excitement is truly inspiring and their discoveries are shaping how we think about the microbiota.

This book would not have been possible without the support of numerous friends and family members. We are indebted to our parents, Dennis and Bonnie Sonnenburg and Aime and Lise Dutil, for too much to comprehensively enumerate. From the extra babysitting to the constant encouragement we are very grateful for their undying support. Finally we need to thank our ultimate inspiration, our two daughters, Claire and Camille. Their willingness to try unusual fermented foods and eat a diet that cares for their microbiota taught us that this next generation can reverse the course of the Western dietary deterioration. We are very proud of their blossoming wisdom that is perfectly captured every time they say they want more kale because their microbiota is hungry . . . and because it's delicious!

APPENDIX

Probiotic-Containing Foods and Drinks

Dairy-based: look for those that are labeled "live and active cultures"

Buttermilk
Crème fraiche
Cultured butter
Cultured cream cheese
Cultured sour cream
Kefir
Lassi
Some cheeses
Yogurt

Vegetable-based†

Kimchee
Pickles
Sauerkraut

Grain- or legume-based†

Miso

Natto
Tempeh

Other

Kombucha—fermented tea
Nondairy probiotic beverages

† Note that heating or cooking will reduce living bacteria.

Daily Dietary Fiber Recommendation

Children	Recommended grams of fiber
1–3 y	19
4–8 y	25

Men	Recommended grams 16of fiber
9–13 y	31
14–18 y	38
19–30 y	38
31–50 y	38
51–70 y	30
> 70 y	30

Women	Recommended grams of fiber
9–13 y	26
14–18 y	26
19–30 y	25
31–50 y	25
51–70 y	21
> 70 y	21
During pregnancy	28
During lactation	29

NOTES

INTRODUCTION

5. **The average American adult:** Yatsunenko, T., et al. "Human Gut Microbiome Viewed across Age and Geography." *Nature* 486.7402 (2012): 222–27. Print.

5. **To top it off we spend an average:** Consumer Expenditures in 2009. U.S. Department of Labor. U.S. Bureau of Labor Statistics. May 2011. Report 1028.

CHAPTER 1: WHAT IS THE MICROBIOTA AND WHY SHOULD I CARE?

10. **For example, within just a few decades:** Robertson, K. L., et al. "Adaptation of the Black Yeast Wangiella Dermatitidis to Ionizing Radiation: Molecular and Cellular Mechanisms." *PLoS One* 7.11 (2012): e48674. Print.

18. **The Hadza microbiota houses:** Schnorr, S. L., et al. "Gut Microbiome of the Hadza Hunter-Gatherers." *Nat Commun* 5 (2014): 3654. Print.

18. **The microbiota from individuals:** Yatsunenko, T., et al. "Human Gut Microbiome Viewed across Age and Geography." *Nature* 486.7402 (2012): 222–27. Print.

18. **Children living in a rural:** De Filippo, C., et al. "Impact of Diet in Shaping Gut Microbiota Revealed by a Comparative Study in Children from Europe and Rural Africa." *Proc Natl Acad Sci U S A* 107.33 (2010): 14691–96. Print. Lin, A., et al. "Distinct Distal Gut Microbiome Diversity and Composition in Healthy Children from Bangladesh and the United States." *PLoS One* 8.1 (2013): e53838. Print.

21. **Nested inside *Tremblaya princeps*:** Husnik, F., et al. "Horizontal Gene Transfer from Diverse Bacteria to an Insect Genome Enables a Tripartite Nested Mealybug Symbiosis." *Cell* 153.7 (2013): 1567–78. Print.
24. **Between the high temperatures:** Thompson, J. D. "The Great Stench or the Fool's Argument." *Yale J Biol Med* 64.5 (1991): 529–41. Print.
25. **In one of the most understated:** Kendall, A. I. "The Bacteria of the Intestinal Tract of Man." *Science* 42.1076 (1915): 209–12. Print.
26. **One of her key discoveries:** Salyers, A. A., et al. "Fermentation of Mucin and Plant Polysaccharides by Strains of Bacteroides from the Human Colon." *Appl Environ Microbiol* 33.2 (1977): 319–22. Print.
29. **When fecal bacteria are compared:** Eckburg, P. B., et al. "Diversity of the Human Intestinal Microbial Flora." *Science* 308.5728 (2005): 1635–38. Print.
31. **Through caring for these mice:** Backhed, F., et al. "The Gut Microbiota as an Environmental Factor That Regulates Fat Storage." *Proc Natl Acad Sci U S A* 101.44 (2004): 15718–23. Print.
31. **They also saw that obese mice:** Ley, R. E., et al. "Obesity Alters Gut Microbial Ecology." *Proc Natl Acad Sci U S A* 102.31 (2005): 11070–75. Print.
32. **Suddenly the lean mice:** Turnbaugh, P. J., et al. "An Obesity-Associated Gut Microbiome with Increased Capacity for Energy Harvest." *Nature* 444.7122 (2006): 1027–31. Print.

CHAPTER 2: ASSEMBLING OUR LIFELONG COMMUNITY OF COMPANIONS

36. **Mice with no microbiota:** Petersson, J., et al. "Importance and Regulation of the Colonic Mucus Barrier in a Mouse Model of Colitis." *Am J Physiol Gastrointest Liver Physiol* 300.2 (2011): G327–33. Print.
38. **The microbiota of C-section babies:** Dominguez-Bello, M. G., et al. "Delivery Mode Shapes the Acquisition and Structure of the Initial Microbiota across Multiple Body Habitats in Newborns." *Proc Natl Acad Sci U S A* 107.26 (2010): 11971–75. Print.
40. **Once the process of necrosis starts:** Lin, P. W., and B. J. Stoll. "Necrotising Enterocolitis." *Lancet* 368.9543 (2006): 1271–83. Print.
40. **While it is unclear which event:** Claud, E. C., et al. "Bacterial Community Structure and Functional Contributions to Emergence of Health or Necrotizing Enterocolitis in Preterm Infants." *Microbiome* 1.1 (2013): 20. Print.

40. **Premature infants already have:** Wang, Y., et al. "16S rRNA Gene-Based Analysis of Fecal Microbiota from Preterm Infants with and without Necrotizing Enterocolitis." *ISME J* 3.8 (2009): 944–54. Print.
40. **Premature infants given beneficial bacteria:** Alfaleh, K., and D. Bassler. "Probiotics for Prevention of Necrotizing Enterocolitis in Preterm Infants." *Cochrane Database Syst Rev.* 1 (2008): Cd005496. Print.
41. **Bacteria like these provide signals:** Tarnow-Mordi, W., and R. F. Soll. "Probiotic Supplementation in Preterm Infants: It Is Time to Change Practice." *J Pediatr* 164.5 (2014): 959–60. Print.
43. **Ruth's team of scientists:** Koren, O., et al. "Host Remodeling of the Gut Microbiome and Metabolic Changes During Pregnancy." *Cell* 150.3 (2012): 470–80. Print.
45. **In 2007 a study:** Palmer, C., et al. "Development of the Human Infant Intestinal Microbiota." *PLoS Biol* 5.7 (2007): e177. Print.
46. **By no accident of nature:** De Filippo, C., et al. "Impact of Diet in Shaping Gut Microbiota Revealed by a Comparative Study in Children from Europe and Rural Africa." *Proc Natl Acad Sci U S A* 107.33 (2010): 14691–96. Print.
46. **But HMOs do more:** Marcobal, A. "Bacteroides in the Infant Gut Consume Milk Oligosaccharides via Mucus-Utilization Pathways." *Cell Host Microbe* 10.5 (2011): 507.14. Print.
47. **Mothers also provide living:** Cabrera-Rubio, R., et al. "The Human Milk Microbiome Changes over Lactation and Is Shaped by Maternal Weight and Mode of Delivery." *Am J Clin Nutr* 96.3 (2012): 544–51. Print.
49. **A group of scientists in the Netherlands:** de Weerth, C., et al. "Intestinal Microbiota of Infants with Colic: Development and Specific Signatures." *Pediatrics* 131.2 (2013): e550–58. Print.
50. **A case study following:** Koenig, J. E., et al. "Succession of Microbial Consortia in the Developing Infant Gut Microbiome." *Proc Natl Acad Sci U S A* 108 Suppl 1 (2011): 4578–85. Print.
56. **Antibiotic use in children:** Trasande, L., et al. "Infant Antibiotic Exposures and Early-Life Body Mass." *Int J Obes* (Lond) 37.1 (2013): 16–23. Print. Hoskin-Parr, L., et al. "Antibiotic Exposure in the First Two Years of Life and Development of Asthma and Other Allergic Diseases by 7.5 Yr: A Dose-Dependent Relationship." *Pediatr Allergy Immunol* 24.8 (2013): 762–71. Print.
56. **Laboratory mice given low doses:** Cho, I., et al. "Antibiotics in Early Life Alter the Murine Colonic Microbiome and Adiposity." *Nature* 488.7413 (2012): 621–26. Print.

57. **A comparison between more than eleven thousand:** Trasande, L., et al. "Infant Antibiotic Exposures and Early-Life Body Mass." *Int J Obes (Lond)* 37.1 (2013): 16–23. Print.

CHAPTER 3: SETTING THE DIAL ON THE IMMUNE SYSTEM

63. **Performing experiments in mice:** Lee, Y. K., et al. "Proinflammatory T-Cell Responses to Gut Microbiota Promote Experimental Autoimmune Encephalomyelitis." *Proc Natl Acad Sci U S A* 108 Suppl 1 (2011): 4615–22. Print.
66. **In 1989, David Strachan:** Strachan, D. P. "Hay Fever, Hygiene, and Household Size." *Bmj* 299.6710 (1989): 1259–60. Print.
67. **The hygiene hypothesis:** Wlasiuk, G., and D. Vercelli. "The Farm Effect, or, When, What and How a Farming Environment Protects from Asthma and Allergic Disease." *Curr Opin Allergy Clin Immunol* 12.5 (2012): 461–66. Print.
68. **Triclosan exposure has recently:** Savage, J. H., et al. "Urinary Levels of Triclosan and Parabens Are Associated with Aeroallergen and Food Sensitization." *J Allergy Clin Immunol* 130.2 (2012): 453–60.e7. Print.
68. **To make matters worse:** Frieden, Thomas. "Antibiotic Resistance and the Threat to Public Health." *Energy and Commerce Subcommittee on Health 2010 of United States House of Representatives.* Print.
69. **But in households with a dog:** Kozyrskyj, A. L., P. Ernst, and A. B. Becker. "Increased Risk of Childhood Asthma from Antibiotic Use in Early Life." *Chest* 131.6 (2007): 1753–59. Print.
69. **Mice that are colonized:** Herbst, T., et al. "Dysregulation of Allergic Airway Inflammation in the Absence of Microbial Colonization." *Am J Respir Crit Care Med* 184.2 (2011): 198–205. Print.
75. **If microbial exposure happens:** Olszak, T., et al. "Microbial Exposure During Early Life Has Persistent Effects on Natural Killer T Cell Function." *Science* 336.6080 (2012): 489–93. Print.
76. **Kenya Honda's research group:** Atarashi, K., et al. "Treg Induction by a Rationally Selected Mixture of Clostridia Strains from the Human Microbiota." *Nature* 500.7461 (2013): 232–36. Print.
77. **These molecules help the intestine:** Smith, P. M., et al. "The Microbial Metabolites, Short-Chain Fatty Acids, Regulate Colonic T_{reg} Homeostasis." *Science* 341.6145 (2013): 569–73. Print.
79. **As studies by Blaser:** Atherton, J. C., and M. J. Blaser. "Coadaptation of Helicobacter Pylori and Humans: Ancient History, Modern Implications." *J Clin Invest* 119.9 (2009): 2475–87. Print.

82. **People who own dogs have bacteria:** Song, S. J., et al. "Cohabiting Family Members Share Microbiota with One Another and with Their Dogs." *Elife* 2 (2013): e00458. Print.

CHAPTER 4: THE TRANSIENTS

88. **The earliest record of fermented food:** McGovern, P. E., et al. "Fermented Beverages of Pre- and Proto-Historic China." *Proc Natl Acad Sci U S A* 101.51 (2004): 17593–98. Print.
89. **In 1908 he recorded:** Metchnikoff, Élie, and P. Chalmers Mitchell. *The Prolongation of Life: Optimistic Studies.* London: Heinemann, 1908. Print.
92. **To test this notion directly:** Merenstein, D., et al. "Use of a Fermented Dairy Probiotic Drink Containing Lactobacillus Casei (DN-114 001) to Decrease the Rate of Illness in Kids: The Drink Study. A Patient-Oriented, Double-Blind, Cluster-Randomized, Placebo-Controlled, Clinical Trial." *Eur J Clin Nutr* 64.7 (2010): 669–77. Print.
93. **A number of studies have shown:** Allen, S. J., et al. "Probiotics for Treating Acute Infectious Diarrhoea." *Cochrane Database Syst Rev.* 11 (2010): Cd003048. Print.
93. **Other trials encompassing thousands:** Hao, Q., et al. "Probiotics for Preventing Acute Upper Respiratory Tract Infections." *Cochrane Database Syst Rev.* 9 (2011): Cd006895. Print.
100. **In many cases the identity:** Sanders, M. E., and J. T. Heimbach. "Functional Foods in the USA: Emphasis on Probiotic Foods." *Food Sci Technol Bull* 1.8 (2004): 1–10. Print.
105. ***Faecalibacterium prausnitzii*:** Cao, Y., J. Shen, and Z. H. Ran. "Association between Faecalibacterium Prausnitzii Reduction and Inflammatory Bowel Disease: A Meta-Analysis and Systematic Review of the Literature." *Gastroenterol Res Pract 2014* (2014): 872725. Print. Fujimoto, T., et al. "Decreased Abundance of Faecalibacterium Prausnitzii in the Gut Microbiota of Crohn's Disease." *J Gastroenterol Hepatol* 28.4 (2013): 613–19. Print. Machiels, K., et al. "A Decrease of the Butyrate-Producing Species Roseburia Hominis and Faecalibacterium Prausnitzii Defines Dysbiosis in Patients with Ulcerative Colitis." *Gut* 63.8 (2014): 1275–83. Print. Balamurugan, R., et al. "Real-Time Polymerase Chain Reaction Quantification of Specific Butyrate-Producing Bacteria, Desulfovibrio and Enterococcus Faecalis in the Feces of Patients with Colorectal Cancer." *J Gastroenterol Hepatol* 23.8 Pt 1 (2008): 1298–303. Print.
105. **Mice harboring this bacterium:** Sokol, H., et al. "Faecalibacterium Prausnitzii Is an Anti-Inflammatory Commensal Bacterium Identified by

Gut Microbiota Analysis of Crohn Disease Patients." *Proc Natl Acad Sci U S A* 105.43 (2008): 16731–36. Print.

106. **In fact, geophagia:** Reid, R. M. "Cultural and Medical Perspectives on Geophagia." *Med Anthropol* 13.4 (1992): 337–51. Print.

106. **There is some evidence that:** Bittner, A. C., R. M. Croffut, and M. C. Stranahan. "Prescript-Assist Probiotic-Prebiotic Treatment for Irritable Bowel Syndrome: A Methodologically Oriented, 2-Week, Randomized, Placebo-Controlled, Double-Blind Clinical Study." *Clin Ther* 27.6 (2005): 755–61. Print.

CHAPTER 5: TRILLIONS OF MOUTHS TO FEED

112. **But rather than "dietary fiber":** Sonnenburg, E. D., and J. L. Sonnenburg. "Starving Our Microbial Self: The Deleterious Consequences of a Diet Deficient in Microbiota-Accessible Carbohydrates." *Cell Metab* (2014). Print.

116. **The list of ways in which:** Russell, W. R., et al. "Colonic Bacterial Metabolites and Human Health." *Curr Opin Microbiol* 16.3 (2013): 246–54. Print.

117. **However, almost one hundred years ago:** Torrey, J. C. "The Regulation of the Intestinal Flora of Dogs through Diet." *J Med Res* 39.3 (1919): 415–47. Print.

118. **In his book *The Saccharine Disease*:** Cleave, T. L. *The Saccharine Disease: Conditions Caused by the Taking of Refined Carbohydrates, Such as Sugar and White Flour.* Keats Publishing, 1975. Print.

119. **The work by Cleave, Burkitt:** Trowell, H. C., and D. P. Burkitt. "The Development of the Concept of Dietary Fibre." *Mol Aspects Med* 9.1 (1987): 7–15. Print.

121. **There are thousands of different types:** Martens, E. C., et al. "The Devil Lies in the Details: How Variations in Polysaccharide Fine-Structure Impact the Physiology and Evolution of Gut Microbes." *J Mol Biol* (2014). Print.

124. **Some of these definitions:** Raninen, K., et al. "Dietary Fiber Type Reflects Physiological Functionality: Comparison of Grain Fiber, Inulin, and Polydextrose." *Nutr Rev* 69.1 (2011): 9–21. Print.

124. **According to the Food and Agriculture Organization:** Dhingra, D., et al. "Dietary Fibre in Foods: A Review." *J Food Sci Technol* 49.3 (2012): 255–66. Print. Westenbrink, S., K. Brunt, and J. W. van der Kamp. "Dietary Fibre: Challenges in Production and Use of Food Composition Data." *Food Chem* 140.3 (2013): 562–67. Print.

125. **During times of low fiber consumption:** Sonnenburg, J. L., et al. "Glycan Foraging in Vivo by an Intestine-Adapted Bacterial Symbiont." *Science* 307.5717 (2005): 1955–59. Print.

125. **While the long-term effects of less gut:** Johansson, M. E., et al. "Bacteria Penetrate the Normally Impenetrable Inner Colon Mucus Layer in Both Murine Colitis Models and Patients with Ulcerative Colitis." *Gut* 63.2 (2014): 281–91. Print.

126. **In 2010 a group of scientists:** Hehemann, J. H., et al. "Bacteria of the Human Gut Microbiome Catabolize Red Seaweed Glycans with Carbohydrate-Active Enzyme Updates from Extrinsic Microbes." *Proc Natl Acad Sci U S A* 109.48 (2012): 19786–91. Print.

128. **In a study published in 2013:** Le Chatelier, E., et al. "Richness of Human Gut Microbiome Correlates with Metabolic Markers." *Nature* 500.7464 (2013): 541–46. Print.

128. **A similar study conducted in France:** Cotillard, A., et al. "Dietary Intervention Impact on Gut Microbial Gene Richness." *Nature* 500.7464 (2013): 585–88. Print.

129. **In 2013 Washington University's:** Ridaura, V. K., et al. "Gut Microbiota from Twins Discordant for Obesity Modulate Metabolism in Mice." *Science* 341.6150 (2013): 12412–14. Print.

133. **But after a few days of training:** Kuoliok, K. E. *Food and Emergency Food in the Circumpolar Area.* Almquist och Wiksell. 1969. Print.

134. **Within four weeks, dieters:** Russell, W. R., et al. "High-Protein, Reduced-Carbohydrate Weight-Loss Diets Promote Metabolite Profiles Likely to Be Detrimental to Colonic Health." *Am J Clin Nutr* 93.5 (2011): 1062–72. Print.

134. **The microbiota of omnivores:** Koeth, R. A., et al. "Intestinal Microbiota Metabolism of L-Carnitine, a Nutrient in Red Meat, Promotes Atherosclerosis." *Nat Med* 19.5 (2013): 576–85. Print.

CHAPTER 6: A GUT FEELING

140. **Scientists noted that microbe-free:** Neufeld, K. M., et al. "Reduced Anxiety-Like Behavior and Central Neurochemical Change in Germ-Free Mice." *Neurogastroenterol Motil* 23.3 (2011): 255–64, e119. Print.

141. **In their observations of these microbe-free:** Diaz Heijtz, R., et al. "Normal Gut Microbiota Modulates Brain Development and Behavior." *Proc Natl Acad Sci U S A* 108.7 (2011): 3047–52. Print.

141. **A group of researchers put two:** Gareau, M. G., et al. "Bacterial Infection Causes Stress-Induced Memory Dysfunction in Mice." *Gut* 60.3 (2011): 307–17. Print.

142. **In 2011 a research group at McMaster University:** Bercik, P., et al. "The Intestinal Microbiota Affect Central Levels of Brain-Derived Neurotropic Factor and Behavior in Mice." *Gastroenterology* 141.2 (2011): 599–609, 09.e1-3. Print.

146. **If the liver fails, these toxic:** Riordan, S. M., and R. Williams. "Gut Flora and Hepatic Encephalopathy in Patients with Cirrhosis." *N Engl J Med* 362.12 (2010): 1140–42. Print.

146. **Before lactulose and rifaximin:** Johnston, G. W., and H. W. Rodgers. "Treatment of Chronic Portal-Systemic Encephalopathy by Colectomy." *Br J Surg* 52 (1965): 424–26. Print.

146. **The kidneys, like the liver:** Aronov, P. A., et al. "Colonic Contribution to Uremic Solutes." *J Am Soc Nephrol* 22.9 (2011): 1769–76. Print.

147. **Researchers at the Cleveland Clinic in Ohio:** Wang, Z., et al. "Gut Flora Metabolism of Phosphatidylcholine Promotes Cardiovascular Disease." *Nature* 472.7341 (2011): 57–63. Print.

147. **In follow-up studies, the researchers found:** Koeth, R. A., et al. "Intestinal Microbiota Metabolism of L-Carnitine, a Nutrient in Red Meat, Promotes Atherosclerosis." *Nat Med* 19.5 (2013): 576–85. Print.

149. **If you induce stress or depression:** O'Mahony, S. M., et al. "Maternal Separation as a Model of Brain-Gut Axis Dysfunction." *Psychopharmacology (Berl)* 214.1 (2011): 71–88. Print.

150. **Maternal separation-induced stress:** O'Mahony, S. M., et al. "Early Life Stress Alters Behavior, Immunity, and Microbiota in Rats: Implications for Irritable Bowel Syndrome and Psychiatric Illnesses." *Biol Psychiatry* 65 (2009): 263–67. Print.

150. **Infant rhesus monkeys that are separated:** Bailey, M. T., and C. L. Coe. "Maternal Separation Disrupts the Integrity of the Intestinal Microflora in Infant Rhesus Monkeys." *Dev Psychobiol* 35.2 (1999): 146–55. Print.

150. **Mice infected with intestinal pathogens:** Lyte, M., et al. "Induction of Anxiety-Like Behavior in Mice During the Initial Stages of Infection with the Agent of Murine Colonic Hyperplasia Citrobacter Rodentium." *Physiol Behav* 89.3 (2006): 350–57. Print. Goehler, L. E., et al. "Campylobacter Jejuni Infection Increases Anxiety-Like Behavior in the Holeboard: Possible Anatomical Substrates for Viscerosensory Modulation of Exploratory Behavior." *Brain Behav Immun* 22.3 (2008): 354–66. Print.

151. **Preliminary studies in humans:** Rao, A. V., et al. "A Randomized, Double-Blind, Placebo-Controlled Pilot Study of a Probiotic in Emotional Symptoms of Chronic Fatigue Syndrome." *Gut Pathog* 1.1 (2009): 6. Print. O'Mahony, L., et al. "Lactobacillus and Bifidobacterium

in Irritable Bowel Syndrome: Symptom Responses and Relationship to Cytokine Profiles." *Gastroenterology* 128.3 (2005): 54–61. Print.

151. **Even healthy volunteers who consumed:** Messaoudi, M., et al. "Assessment of Psychotropic-Like Properties of a Probiotic Formulation (Lactobacillus Helveticus R0052 and Bifidobacterium Longum R0175) in Rats and Human Subjects." *Br J Nutr* 105.5 (2011): 755–64. Print.
152. **While marked differences have been:** Cao, X., et al. "Characteristics of the Gastrointestinal Microbiome in Children with Autism Spectrum Disorder: A Systematic Review." *Shanghai Arch Psychiatry* 25.6 (2013): 342–53. Print.
153. **In 2013, a group of scientists from Caltech:** Hsiao, E. Y., et al. "Microbiota Modulate Behavioral and Physiological Abnormalities Associated with Neurodevelopmental Disorders." *Cell* 155.7 (2013): 1451–63. Print.
156. **In 2013, a group of scientists at UCLA:** Tillisch, K., et al. "Consumption of Fermented Milk Product with Probiotic Modulates Brain Activity." *Gastroenterology* 144.7 (2013): 1394–401, 401.e1-4. Print.
161. **"How these differences in our microbial":** Insel, Thomas. "The Top Ten Research Advances of 2012." National Institute of Mental Health Director's Blog 2012. Web.

CHAPTER 7: EAT SH!T AND LIVE

164. **Collectively, Americans experience approximately:** DuPont, Herbert L. "Acute Infectious Diarrhea in Immunocompetent Adults." *New Engl J Med* 370.16 (2014): 1532.
166. **CDAD is responsible for the death of approximately:** McDonald, L. C. et al. "Vital Signs: Preventing *Clostridium difficile* Infections." MMWR Morb Mortal Wkly Rep 61.9 (2012): 157–62. Print.
166. *In a hospital setting that number rises*: Goudarzi, M., et al. "*Clostridium difficile* Infection: Epidemiology, Pathogenesis, Risk Factors, and Therapeutic Options." *Scientifica* 2014 (2014): 916826. Print.
167. **In 2013, a group of scientists and physicians:** van Nood, E., et al. "Duodenal Infusion of Donor Feces for Recurrent Clostridium Difficile." *N Engl J Med* 368.5 (2013): 407–15. Print.
168. **In 1958, Dr. Ben Eiseman:** Eiseman, B., et al. "Fecal Enema as an Adjunct in the Treatment of Pseudomembranous Enterocolitis." *Surgery* 44.5 (1958): 854–59. Print.
169. **From fourth-century China:** Zhang, F., et al. "Should We Standardize the 1,700-Year-Old Fecal Microbiota Transplantation?" *Am J Gastroenterol* 107.11 (2012): 1755; author reply pp. 55–56. Print.

171. **David Relman and Les Dethlefsen:** Dethlefsen, L., and D. A. Relman. "Incomplete Recovery and Individualized Responses of the Human Distal Gut Microbiota to Repeated Antibiotic Perturbation." *Proc Natl Acad Sci U S A* 108 Suppl 1 (2011): 4554-61. Print.

174. **In a healthy, colonization-resistant microbiota:** Ng, K. M., et al. "Microbiota-Liberated Host Sugars Facilitate Post-Antibiotic Expansion of Enteric Pathogens." *Nature* 502.7469 (2013): 96–99. Print.

177. **He found that both diarrhea:** Kashyap, P. C., et al. "Complex Interactions among Diet, Gastrointestinal Transit, and Gut Microbiota in Humanized Mice." *Gastroenterology* 144.5 (2013): 967–77. Print.

178. **Most physicians do conduct safety:** Smith, M. B., C. Kelly, and E. J. Alm. "Policy: How to Regulate Faecal Transplants." *Nature* 506.7488 (2014): 290–1. Print.

179. **A small-scale clinical trial investigating:** Vrieze, A., et al. "Transfer of Intestinal Microbiota from Lean Donors Increases Insulin Sensitivity in Individuals with Metabolic Syndrome." *Gastroenterology* 143.4 (2012): 913-6.e7. Print.

179. **Treating inflammatory bowel disease:** van Nood, E., et al. "Fecal Microbiota Transplantation: Facts and Controversies." *Curr Opin Gastroenterol* 30.1 (2014): 34–39. Print.

181. **A group of scientists has created:** Petrof, E. O., et al. "Stool Substitute Transplant Therapy for the Eradication of Clostridium Difficile Infection: 'Repoopulating' the Gut." *Microbiome* 1.1 (2013): 3. Print.

184. **Early reports indicate that:** Vrieze, A., et al. "Transfer of Intestinal Microbiota from Lean Donors Increases Insulin Sensitivity in Individuals with Metabolic Syndrome." *Gastroenterology* 143.4 (2012): 913–6.e7. Print. Nieuwdorp, M., A. Vrieze, and W. M. de Vos. "Reply to Konstantinov and Peppelenbosch." *Gastroenterology* 144.4 (2013): e20-1. Print.

184. **Antibiotics are the usual course of action:** Rabbani, G. H., et al. "Green Banana Reduces Clinical Severity of Childhood Shigellosis: A Double-Blind, Randomized, Controlled Clinical Trial." *Pediatr Infect Dis J* 28.5 (2009): 420–25. Print.

CHAPTER 8: THE AGING MICROBIOTA

188. **Regardless of the minor disturbances:** Faith, J. J., et al. "The Long-Term Stability of the Human Gut Microbiota." *Science* 341.6141 (2013): 1237439. Print.

188–89. **It seems as if some species of bacteria:** Lee, S. M., et al. "Bacterial Colonization Factors Control Specificity and Stability of the Gut Microbiota." *Nature* 501.7467 (2013): 426–29. Print.

190. **In 2007, a group of researchers:** Claesson, M. J., et al. "Gut Microbiota Composition Correlates with Diet and Health in the Elderly." *Nature* 488.7410 (2012): 178–84. Print.

192. **Similar studies performed in elderly:** Mueller, S., et al. "Differences in Fecal Microbiota in Different European Study Populations in Relation to Age, Gender, and Country: A Cross-Sectional Study." *Appl Environ Microbiol* 72.2 (2006): 1027–33. Print.

194. **Laboratory mice fed a diet rich:** Devkota, S., et al. "Dietary-Fat-Induced Taurocholic Acid Promotes Pathobiont Expansion and Colitis in Il10-/- Mice." *Nature* 487.7405 (2012): 104–8. Print.

195. **Studies in laboratory mice, which allow:** Evans, C. C., et al. "Exercise Prevents Weight Gain and Alters the Gut Microbiota in a Mouse Model of High Fat Diet-Induced Obesity." *PLoS One* 9.3 (2014): e92193. Print.

196. **Scientists studying mice that were:** Viaud, S., et al. "The Intestinal Microbiota Modulates the Anticancer Immune Effects of Cyclophosphamide." *Science* 342.6161 (2013): 971–76. Print.

197. **A group of researchers investigated:** Iida, N., et al. "Commensal Bacteria Control Cancer Response to Therapy by Modulating the Tumor Microenvironment." *Science* 342.6161 (2013): 967–70. Print.

200. **Acetaminophen overdose is the leading:** Fontana, R. J. "Acute Liver Failure including Acetaminophen Overdose." *Med Clin North Am.* 92.2 (2008): 761–94. Print.

201. **When studying how quickly acetaminophen:** Clayton, T. A., et al. "Pharmacometabonomic Identification of a Significant Host-Microbiome Metabolic Interaction Affecting Human Drug Metabolism." *Proc Natl Acad Sci U S A* 106.34 (2009): 14728–33. Print.

202. **Vincent van Gogh is thought to have:** Wolf, P. "Creativity and Chronic Disease: Vincent Van Gogh (1853–1890)." *West J Med* 175.5 (2001): 348. Print.

203. **Mice with *Eggerthella lenta* eating a high-protein diet:** Haiser, H. J., et al. "Predicting and Manipulating Cardiac Drug Inactivation by the Human Gut Bacterium Eggerthella Lenta." *Science* 341.6143 (2013): 295–98. Print.

204. **Human examples of extreme longevity:** Biagi, E., et al. "Through Ageing, and Beyond: Gut Microbiota and Inflammatory Status in Seniors and Centenarians." *PLoS One* 5.5 (2010): e10667. Print.

205. **In fact, increased fiber consumption among:** Cuervo, A., et al. "Fiber from a Regular Diet Is Directly Associated with Fecal Short-Chain Fatty Acid Concentrations in the Elderly." *Nutr Res* 33.10 (2013): 811–16. Print.

CHAPTER 9: MANAGING YOUR INTERNAL FERMENTATION

210. **But in fact this is not the case:** Turnbaugh, P. J., et al. "A Core Gut Microbiome in Obese and Lean Twins. *Nature* 457.7228 (2009): 480–84.

212. **Although admittedly a lot of work:** Ip, S., et al. "Breastfeeding and Maternal and Infant Health Outcomes in Developed Countries." *Evid Rep Technol Assess* (Full Rep). 153 (2007): 1–186. Print.

214. **Children raised on farms are much:** Wlasiuk, G., and D. Vercelli. "The Farm Effect, or, When, What and How a Farming Environment Protects from Asthma and Allergic Disease." *Curr Opin Allergy Clin Immunol* 12.5 (2012): 461–66. Print.

215. **A recent study found that children:** Hesselmar, B., et al. "Pacifier Cleaning Practices and Risk of Allergy Development." *Pediatrics* 131.6 (2013): e1829–37. Print.

218. **Obese and overweight individuals:** Cotillard, A., et al. "Dietary Intervention Impact on Gut Microbial Gene Richness." *Nature* 500.7464 (2013): 585–88. Print.

BIBLIOGRAPHY

Alfaleh, K., and D. Bassler. "Probiotics for Prevention of Necrotizing Enterocolitis in Preterm Infants." *Cochrane Database Syst Rev.* 1 (2008): Cd005496. Print.

Allen, S. J., et al. "Probiotics for Treating Acute Infectious Diarrhoea." *Cochrane Database Syst Rev.* 11 (2010): Cd003048. Print.

Alvarez-Acosta, T., et al. "Beneficial Role of Green Plantain [Musa paradisiaca] in the Management of Persistent Diarrhea: A Prospective Randomized Trial." *J Am Coll Nutr* 28.2 (2009): 169–76. Print.

Aronov, P. A., et al. "Colonic Contribution to Uremic Solutes." *J Am Soc Nephrol* 22.9 (2011): 1769–76. Print.

Atarashi, K., et al. "Treg Induction by a Rationally Selected Mixture of Clostridia Strains from the Human Microbiota." *Nature* 500.7461 (2013): 232–36. Print.

Atherton, J. C., and M. J. Blaser. "Coadaptation of Helicobacter Pylori and Humans: Ancient History, Modern Implications." *J Clin Invest* 119.9 (2009): 2475–87. Print.

Backhed, F., et al. "The Gut Microbiota as an Environmental Factor That Regulates Fat Storage." *Proc Natl Acad Sci U S A* 101.44 (2004): 15718–23. Print.

Bailey, M. T., and C. L. Coe. "Maternal Separation Disrupts the Integrity of the Intestinal Microflora in Infant Rhesus Monkeys." *Dev Psychobiol* 35.2 (1999): 146–55. Print.

Balamurugan, R., et al. "Real-Time Polymerase Chain Reaction Quantification of Specific Butyrate-Producing Bacteria, Desulfovibrio and Enterococcus Faecalis in the Feces of Patients with Colorectal Cancer." *J Gastroenterol Hepatol* 23.8 Pt 1 (2008): 1298–303. Print.

Bercik, P., et al. "The Intestinal Microbiota Affect Central Levels of Brain-Derived Neurotropic Factor and Behavior in Mice." *Gastroenterology* 141.2 (2011): 599–609, 09.e1-3. Print.

Biagi, E., et al. "Through Ageing, and Beyond: Gut Microbiota and Inflammatory Status in Seniors and Centenarians." *PLoS One* 5.5 (2010): e10667. Print.

Bittner, A. C., R. M. Croffut, and M. C. Stranahan. "Prescript-Assist Probiotic-Prebiotic Treatment for Irritable Bowel Syndrome: A Methodologically Oriented, 2-Week, Randomized, Placebo-Controlled, Double-Blind Clinical Study." *Clin Ther* 27.6 (2005): 755–61. Print.

Cabrera-Rubio, R., et al. "The Human Milk Microbiome Changes over Lactation and Is Shaped by Maternal Weight and Mode of Delivery." *Am J Clin Nutr* 96.3 (2012): 544–51. Print.

Cao, X., et al. "Characteristics of the Gastrointestinal Microbiome in Children with Autism Spectrum Disorder: A Systematic Review." *Shanghai Arch Psychiatry* 25.6 (2013): 342–53. Print.

Cao, Y., J. Shen, and Z. H. Ran. "Association between Faecalibacterium Prausnitzii Reduction and Inflammatory Bowel Disease: A Meta-Analysis and Systematic Review of the Literature." *Gastroenterol Res Pract* 2014 (2014): 872725. Print.

Cho, I., et al. "Antibiotics in Early Life Alter the Murine Colonic Microbiome and Adiposity." *Nature* 488.7413 (2012): 621–26. Print.

Claesson, M. J., et al. "Gut Microbiota Composition Correlates with Diet and Health in the Elderly." *Nature* 488.7410 (2012): 178–84. Print.

Claud, E. C., et al. "Bacterial Community Structure and Functional Contributions to Emergence of Health or Necrotizing Enterocolitis in Preterm Infants." *Microbiome* 1.1 (2013): 20. Print.

Clayton, T. A., et al. "Pharmacometabonomic Identification of a Significant Host-Microbiome Metabolic Interaction Affecting Human Drug Metabolism." *Proc Natl Acad Sci U S A* 106.34 (2009): 14728–33. Print.

Cleave, T. L. *The Saccharine Disease: Conditions Caused by the Taking of Refined Carbohydrates, Such as Sugar and White Flour.* Keats Publishing, 1975. Print.

Cotillard, A., et al. "Dietary Intervention Impact on Gut Microbial Gene Richness." *Nature* 500.7464 (2013): 585–88. Print.

Cuervo, A., et al. "Fiber from a Regular Diet Is Directly Associated with Fecal Short-Chain Fatty Acid Concentrations in the Elderly." *Nutr Res* 33.10 (2013): 811–16. Print.

De Filippo, C., et al. "Impact of Diet in Shaping Gut Microbiota Revealed by a Comparative Study in Children from Europe and Rural Africa." *Proc Natl Acad Sci U S A* 107.33 (2010): 14691–96. Print.

de Weerth, C., et al. "Intestinal Microbiota of Infants with Colic: Development and Specific Signatures." *Pediatrics* 131.2 (2013): e550–8. Print.

Dethlefsen, L., et al. "The Pervasive Effects of an Antibiotic on the Human Gut Microbiota, as Revealed by Deep 16S rRNA Sequencing." *PLoS Biol* 6.11 (2008): e280. Print.

Dethlefsen, L., and D. A. Relman. "Incomplete Recovery and Individualized Responses of the Human Distal Gut Microbiota to Repeated Antibiotic Perturbation." *Proc Natl Acad Sci U S A* 108 Suppl 1 (2011): 4554–61. Print.

Devkota, S., et al. "Dietary-Fat-Induced Taurocholic Acid Promotes Pathobiont Expansion and Colitis in Il10-/- Mice." *Nature* 487.7405 (2012): 104–8. Print.

Dhingra, D., et al. "Dietary Fibre in Foods: A Review." *J Food Sci Technol* 49.3 (2012): 255–66. Print.

Diaz Heijtz, R., et al. "Normal Gut Microbiota Modulates Brain Development and Behavior." *Proc Natl Acad Sci U S A* 108.7 (2011): 3047–52. Print.

Dominguez-Bello, M. G., et al. "Delivery Mode Shapes the Acquisition and Structure of the Initial Microbiota across Multiple Body Habitats in Newborns." *Proc Natl Acad Sci U S A* 107.26 (2010): 11971–75. Print.

Eckburg, P. B., et al. "Diversity of the Human Intestinal Microbial Flora." *Science* 308.5728 (2005): 1635–38. Print.

Eiseman, B., et al. "Fecal Enema as an Adjunct in the Treatment of Pseudomembranous Enterocolitis." *Surgery* 44.5 (1958): 854–59. Print.

Evans, C. C., et al. "Exercise Prevents Weight Gain and Alters the Gut Microbiota in a Mouse Model of High Fat Diet-Induced Obesity." *PLoS One* 9.3 (2014): e92193. Print.

Faith, J. J., et al. "The Long-Term Stability of the Human Gut Microbiota." *Science* 341.6141 (2013): 1237439. Print.

Fontana, R. J. "Acute Liver Failure Including Acetaminophen Overdose." *Med Clin North Am.* 92.2 (2008): 761–94. Print.

Frieden, Thomas. "Antibiotic Resistance and the Threat to Public Health." *Energy and Commerce Subcommittee on Health 2010 of United States House of Representatives.* Print.

Fujimoto, T., et al. "Decreased Abundance of Faecalibacterium prausnitzii in the Gut Microbiota of Crohn's Disease." *J Gastroenterol Hepatol* 28.4 (2013): 613–19. Print.

Gareau, M. G., et al. "Bacterial Infection Causes Stress-Induced Memory Dysfunction in Mice." *Gut* 60.3 (2011): 307–17. Print.

Goehler, L. E., et al. "Campylobacter Jejuni Infection Increases Anxiety-Like Behavior in the Holeboard: Possible Anatomical Substrates for Viscerosensory Modulation of Exploratory Behavior." *Brain Behav Immun* 22.3 (2008): 354–66. Print.

Goudarzi, M., et al. "Clostridium difficile Infection: Epidemiology, Pathogenesis, Risk Factors, and Therapeutic Options." *Scientifica* 2014 (2014): 916826. Print.

Haiser, H. J., et al. "Predicting and Manipulating Cardiac Drug Inactivation by the Human Gut Bacterium Eggerthella Lenta." *Science* 341.6143 (2013): 295–98. Print.

Hao, Q., et al. "Probiotics for Preventing Acute Upper Respiratory Tract Infections." *Cochrane Database Syst Rev.* 9 (2011): Cd006895. Print.

Hehemann, J. H., et al. "Bacteria of the Human Gut Microbiome Catabolize Red Seaweed Glycans with Carbohydrate-Active Enzyme Updates from Extrinsic Microbes." *Proc Natl Acad Sci U S A* 109.48 (2012): 19786–91. Print.

Herbst, T., et al. "Dysregulation of Allergic Airway Inflammation in the Absence of Microbial Colonization." *Am J Respir Crit Care Med* 184.2 (2011): 198–205. Print.

Hesselmar, B., et al. "Pacifier Cleaning Practices and Risk of Allergy Development." *Pediatrics* 131.6 (2013): e1829–37. Print.

Hoskin-Parr, L., et al. "Antibiotic Exposure in the First Two Years of Life and Development of Asthma and Other Allergic Diseases by 7.5 Yr: A Dose-Dependent Relationship." *Pediatr Allergy Immunol* 24.8 (2013): 762–71. Print.

Hsiao, E. Y., et al. "Microbiota Modulate Behavioral and Physiological Abnormalities Associated with Neurodevelopmental Disorders." *Cell* 155.7 (2013): 1451–63. Print.

Husnik, F., et al. "Horizontal Gene Transfer from Diverse Bacteria to an Insect Genome Enables a Tripartite Nested Mealybug Symbiosis." *Cell* 153.7 (2013): 1567–78. Print.

Iida, N., et al. "Commensal Bacteria Control Cancer Response to Therapy by Modulating the Tumor Microenvironment." *Science* 342.6161 (2013): 967–70. Print.

Insel, Thomas. "The Top Ten Research Advances of 2012." National Institute of Mental Health Director's Blog 2012. Web.

Ip, S., et al. "Breastfeeding and Maternal and Infant Health Outcomes in Developed Countries." *Evid Rep Technol Assess (Full Rep)* 153 (2007): 1–186. Print.

Johansson, M. E., et al. "Bacteria Penetrate the Normally Impenetrable Inner Colon Mucus Layer in Both Murine Colitis Models and Patients with Ulcerative Colitis." *Gut* 63.2 (2014): 281–91. Print.

Johnston, G. W., and H. W. Rodgers. "Treatment of Chronic Portal-Systemic Encephalopathy by Colectomy." *Br J Surg* 52 (1965): 424–26. Print.

Kashyap, P. C., et al. "Complex Interactions among Diet, Gastrointestinal Transit, and Gut Microbiota in Humanized Mice." *Gastroenterology* 144.5 (2013): 967–77. Print.

Kendall, A. I. "The Bacteria of the Intestinal Tract of Man." *Science* 42.1076 (1915): 209–12. Print.

Koenig, J. E., et al. "Succession of Microbial Consortia in the Developing Infant Gut Microbiome." *Proc Natl Acad Sci U S A* 108 Suppl 1 (2011): 4578–85. Print.

Koeth, R. A., et al. "Intestinal Microbiota Metabolism of L-Carnitine, a Nutrient in Red Meat, Promotes Atherosclerosis." *Nat Med* 19.5 (2013): 576–85. Print.

Koren, O., et al. "Host Remodeling of the Gut Microbiome and Metabolic Changes During Pregnancy." *Cell* 150.3 (2012): 470–80. Print.

Kozyrskyj, A. L., P. Ernst, and A. B. Becker. "Increased Risk of Childhood Asthma from Antibiotic Use in Early Life." *Chest* 131.6 (2007): 1753–59. Print.

Kuoliok, K. E. *Food and Emergency Food in the Circumpolar Area.* Almquist och Wiksell, 1969. Print.

Le Chatelier, E., et al. "Richness of Human Gut Microbiome Correlates with Metabolic Markers." *Nature* 500.7464 (2013): 541–46. Print.

Lee, S. M., et al. "Bacterial Colonization Factors Control Specificity and Stability of the Gut Microbiota." *Nature* 501.7467 (2013): 426–29. Print.

Lee, Y. K., et al. "Proinflammatory T-Cell Responses to Gut Microbiota Promote Experimental Autoimmune Encephalomyelitis." *Proc Natl Acad Sci U S A* 108 Suppl 1 (2011): 4615–22. Print.

Lewis, S. J., and K. W. Heaton. "Stool Form Scale as a Useful Guide to Intestinal Transit Time." *Scand J Gastroenterol* 32.9 (1997): 920–24. Print.

Ley, R. E., et al. "Obesity Alters Gut Microbial Ecology." *Proc Natl Acad Sci U S A* 102.31 (2005): 11070–75. Print.

Lin, A., et al. "Distinct Distal Gut Microbiome Diversity and Composition in Healthy Children from Bangladesh and the United States." *PLoS One* 8.1 (2013): e53838. Print.

Lin, P. W., and B. J. Stoll. "Necrotising Enterocolitis." *Lancet* 368.9543 (2006): 1271–83. Print.

Lyte, M., et al. "Induction of Anxiety-Like Behavior in Mice During the Initial Stages of Infection with the Agent of Murine Colonic Hyperplasia Citrobacter Rodentium." *Physiol Behav* 89.3 (2006): 350–57. Print.

Machiels, K., et al. "A Decrease of the Butyrate-Producing Species Roseburia Hominis and Faecalibacterium Prausnitzii Defines Dysbiosis in Patients with Ulcerative Colitis." *Gut* 63.8 (2014): 1275–83. Print.

Marcobal, A., "Bacteroides in the Infant Gut Consume Milk Oligosaccharides via Mucus-Utilization Pathways." *Cell Host Microbe* 10.5 (2011): 507–14. Print.

Martens, E. C., et al. "The Devil Lies in the Details: How Variations in Polysaccharide Fine-Structure Impact the Physiology and Evolution of Gut Microbes." *J Mol Biol* (2014). Print.

McDonald, L. C., et al. "Vital Signs: Preventing Clostridium difficile Infections. MMWR Morb Mortal Wkly Rep 61.9 (2012): 1157–67. Print.

McGovern, P. E., et al. "Fermented Beverages of Pre- and Proto-Historic China." *Proc Natl Acad Sci U S A* 101.51 (2004): 17593–98. Print.

Merenstein, D., et al. "Use of a Fermented Dairy Probiotic Drink Containing Lactobacillus Casei (DN-114 001) to Decrease the Rate of Illness in Kids: The Drink Study. A Patient-Oriented, Double-Blind, Cluster-Randomized, Placebo-Controlled, Clinical Trial." *Eur J Clin Nutr* 64.7 (2010): 669–77. Print.

Messaoudi, M., et al. "Assessment of Psychotropic-Like Properties of a Probiotic Formulation (Lactobacillus Helveticus R0052 and Bifidobacterium Longum R0175) in Rats and Human Subjects." *Br J Nutr* 105.5 (2011): 755–64. Print.

Metchnikoff, Élie, and P. Chalmers Mitchell. *The Prolongation of Life: Optimistic Studies*. London: Heinemann, 1908. Print.

Mueller, S., et al. "Differences in Fecal Microbiota in Different European Study Populations in Relation to Age, Gender, and Country: A Cross-Sectional Study." *Appl Environ Microbiol* 72.2 (2006): 1027–33. Print.

Neufeld, K. M., et al. "Reduced Anxiety-Like Behavior and Central Neurochemical Change in Germ-Free Mice." *Neurogastroenterol Motil* 23.3 (2011): 255–64, e119. Print.

Ng, K. M., et al. "Microbiota-Liberated Host Sugars Facilitate Post-Antibiotic Expansion of Enteric Pathogens." *Nature* 502.7469 (2013): 96–99. Print.

Nieuwdorp, M., A. Vrieze, and W. M. de Vos. "Reply to Konstantinov and Peppelenbosch." *Gastroenterology* 144.4 (2013): e20–21. Print.

Olszak, T., et al. "Microbial Exposure During Early Life Has Persistent Effects on Natural Killer T Cell Function." *Science* 336.6080 (2012): 489–93. Print.

O'Mahony, L., et al. "Lactobacillus and Bifidobacterium in Irritable Bowel Syndrome: Symptom Responses and Relationship to Cytokine Profiles." *Gastroenterology* 128.3 (2005): 541–51. Print.

O'Mahony, S. M., et al. "Maternal Separation as a Model of Brain-Gut Axis Dysfunction." *Psychopharmacology (Berl)* 214.1 (2011): 71–88. Print.

Palmer, C., et al. "Development of the Human Infant Intestinal Microbiota." *PLoS Biol* 5.7 (2007): e177. Print.

Petersson, J., et al. "Importance and Regulation of the Colonic Mucus Barrier in a Mouse Model of Colitis." *Am J Physiol Gastrointest Liver Physiol* 300.2 (2011): G327–33. Print.

Petrof, E. O., et al. "Stool Substitute Transplant Therapy for the Eradication of Clostridium Difficile Infection: 'Repoopulating' the Gut." *Microbiome* 1.1 (2013): 3. Print.

Rabbani, G. H., et al. "Green Banana Reduces Clinical Severity of Childhood Shigellosis: A Double-Blind, Randomized, Controlled Clinical Trial." *Pediatr Infect Dis J* 28.5 (2009): 420–25. Print.

Raninen, K., et al. "Dietary Fiber Type Reflects Physiological Functionality: Comparison of Grain Fiber, Inulin, and Polydextrose." *Nutr Rev* 69.1 (2011): 9–21. Print.

Rao, A. V., et al. "A Randomized, Double-Blind, Placebo-Controlled Pilot Study of a Probiotic in Emotional Symptoms of Chronic Fatigue Syndrome." *Gut Pathog* 1.1 (2009): 6. Print.

Reid, R. M. "Cultural and Medical Perspectives on Geophagia." *Med Anthropol* 13.4 (1992): 337–51. Print.

Ridaura, V. K., et al. "Gut Microbiota from Twins Discordant for Obesity Modulate Metabolism in Mice." *Science* 341.6150 (2013): 1241214. Print.

Riordan, S. M., and R. Williams. "Gut Flora and Hepatic Encephalopathy in Patients with Cirrhosis." *N Engl J Med* 362.12 (2010): 1140–42. Print.

Robertson, K. L., et al. "Adaptation of the Black Yeast Wangiella Dermatitidis to Ionizing Radiation: Molecular and Cellular Mechanisms." *PLoS One* 7.11 (2012): e48674. Print.

Russell, W. R., et al. "High-Protein, Reduced-Carbohydrate Weight-Loss Diets Promote Metabolite Profiles Likely to Be Detrimental to Colonic Health." *Am J Clin Nutr* 93.5 (2011): 1062–72. Print.

Russell, W. R., et al. "Colonic Bacterial Metabolites and Human Health." *Curr Opin Microbiol* 16.3 (2013): 246–54. Print.

Salyers, A. A., et al. "Fermentation of Mucin and Plant Polysaccharides by Strains of Bacteroides from the Human Colon." *Appl Environ Microbiol* 33.2 (1977): 319–22. Print.

Sanders, M. E., and J. T. Heimbach. "Functional Foods in the USA: Emphasis on Probiotic Foods." *Food Sci Technol Bull* 1.8 (2004): 1–10. Print.

Savage, J. H., et al. "Urinary Levels of Triclosan and Parabens Are Associated with Aeroallergen and Food Sensitization." *J Allergy Clin Immunol* 130.2 (2012): 453–60.e7. Print.

Schnorr, S. L., et al. "Gut Microbiome of the Hadza Hunter-Gatherers." *Nat Commun* 5 (2014): 3654. Print.

Smith, M. B., C. Kelly, and E. J. Alm. "Policy: How to Regulate Faecal Transplants." *Nature* 506.7488 (2014): 290–91. Print.

Smith, P. M., et al. "The Microbial Metabolites, Short-Chain Fatty Acids, Regulate Colonic Treg Homeostasis." *Science* 341.6145 (2013): 569–73. Print.

Sokol, H., et al. "Faecalibacterium Prausnitzii Is an Anti-Inflammatory Commensal Bacterium Identified by Gut Microbiota Analysis of Crohn Disease Patients." *Proc Natl Acad Sci U S A* 105.43 (2008): 16731–36. Print.

Song, S. J., et al. "Cohabiting Family Members Share Microbiota with One Another and with Their Dogs." *Elife* 2 (2013): e00458. Print.

Sonnenburg, E. D., and J. L. Sonnenburg. "Starving Our Microbial Self: The Deleterious Consequences of a Diet Deficient in Microbiota-Accessible Carbohydrates." *Cell Metab* (2014). Print.

Sonnenburg, J. L., et al. "Glycan Foraging in Vivo by an Intestine-Adapted Bacterial Symbiont." *Science* 307.5717 (2005): 1955–59. Print.

Strachan, D. P. "Hay Fever, Hygiene, and Household Size." *Bmj* 299.6710 (1989): 1259–60. Print.

Sudo, N., et al. "Postnatal Microbial Colonization Programs the Hypothalamic-Pituitary-Adrenal System for Stress Response in Mice." *J Physiol* 558.Pt 1 (2004): 263–75. Print.

Tarnow-Mordi, W., and R. F. Soll. "Probiotic Supplementation in Preterm Infants: It Is Time to Change Practice." *J Pediatr* 164.5 (2014): 959–60. Print.

Thompson, J. D. "The Great Stench or the Fool's Argument." *Yale J Biol Med* 64.5 (1991): 529–41. Print.

Tillisch, K., et al. "Consumption of Fermented Milk Product with Probiotic Modulates Brain Activity." *Gastroenterology* 144.7 (2013): 1394–401, 401.e1–4. Print.

Torrey, J. C. "The Regulation of the Intestinal Flora of Dogs through Diet." *J Med Res* 39.3 (1919): 415–47. Print.

Trasande, L., et al. "Infant Antibiotic Exposures and Early-Life Body Mass." *Int J Obes (Lond)* 37.1 (2013): 16–23. Print.

Trowell, H. C., and D. P. Burkitt. "The Development of the Concept of Dietary Fibre." *Mol Aspects Med* 9.1 (1987): 7–15. Print.

Turnbaugh, P. J., et al. "An Obesity-Associated Gut Microbiome with Increased Capacity for Energy Harvest." *Nature* 444.7122 (2006): 1027–31. Print.

van Nood, E., et al. "Duodenal Infusion of Donor Feces for Recurrent Clostridium Difficile." *N Engl J Med* 368.5 (2013): 407–15. Print.

van Nood, E., et al. "Fecal Microbiota Transplantation: Facts and Controversies." *Curr Opin Gastroenterol* 30.1 (2014): 34–39. Print.

Viaud, S., et al. "The Intestinal Microbiota Modulates the Anticancer Immune Effects of Cyclophosphamide." *Science* 342.6161 (2013): 971–76. Print.

Vrieze, A., et al. "Transfer of Intestinal Microbiota from Lean Donors Increases Insulin Sensitivity in Individuals with Metabolic Syndrome." *Gastroenterology* 143.4 (2012): 913–6.e7. Print.

Wang, Y., et al. "16S rRNA Gene-Based Analysis of Fecal Microbiota from Preterm Infants with and without Necrotizing Enterocolitis." *ISME J* 3.8 (2009): 944–54. Print.

Wang, Z., et al. "Gut Flora Metabolism of Phosphatidylcholine Promotes Cardiovascular Disease." *Nature* 472.7341 (2011): 57–63. Print.

Westenbrink, S., K. Brunt, and J. W. van der Kamp. "Dietary Fibre: Challenges in Production and Use of Food Composition Data." *Food Chem* 140.3 (2013): 562–67. Print.

Wlasiuk, G., and D. Vercelli. "The Farm Effect, or, When, What and How a Farming Environment Protects from Asthma and Allergic Disease." *Curr Opin Allergy Clin Immunol* 12.5 (2012): 461–66. Print.

Wolf, P. "Creativity and Chronic Disease. Vincent van Gogh (1853-1890)." *West J Med* 175.5 (2001): 348. Print.

Yatsunenko, T., et al. "Human Gut Microbiome Viewed across Age and Geography." *Nature* 486.7402 (2012): 222–27. Print.

Zhang, F., et al. "Should We Standardize the 1,700-Year-Old Fecal Microbiota Transplantation?" *Am J Gastroenterol* 107.11 (2012): 1755; author reply pp. 55–56. Print.

INDEX

acetaminophen, 200, 201–2
aging, 4, 187–89, 204–7
  and bacterial niches or "professions," 189
  centenarians, 204
  and diet, 190–92, 204–5
  and ELDERMET, 190–92
  environmental shifts in, 189–93
  health decline in, 203–4
  and hereditary microbiota, 188
  and inflammaging, 193–95, 203, 205
  and medications, 200–203
agriculture:
  birth of, 15–16, 111
  CSAs, 214
allergies, 61, 79, 211, 214
Almond/Parsley Pesto, 256
Almond/Walnut Butter, 240–41
American Gut Project, 30, 38, 224–25
amino acids, 202–3
anaerobic environment, 114–15
anthrax, 24
antibiotics:
  broad-spectrum, 55, 169–73
  and chemotherapy, 197, 199
  for children, 56, 57, 58, 69, 212–13
  development of, 25
  diarrhea and inflammation caused by, 59, 96, 108–9, 165–67, 175
  effects on microbiota, 55–56, 58–59, 67, 170–77, 212–13
  fed to livestock, 56–57
  overuse of, 2, 5, 19, 160, 212–13
  probiotics following a course of, 108, 221
  superbugs resistant to, 81, 170, 176–77, 213
antibodies, 78
anxiety, 150–51, 160
appendices:
  daily fiber recommendation, 270
  feeding your microbiota, 230–31
  fermented foods, 269–70
  menus, 227–29
  recipes, 232–65
arabinoxylan, 126
archaea, 9
arginine, 203
asthma, 69, 79, 211, 214
atherosclerosis, 218
autism, 32, 152–56, 160
autoimmune disease, 2, 5–6, 61, 67, 69, 76
autonomic nervous system, 138
avocados, 219, 237–38, 242
Aztec Hot Chocolate, 238–39

babies, *see* infants
baby formulas, 47–48
bacteria:
  antibiotics vs., 169–73
  colonies of, 225
  disease-causing, 25
  everywhere, 10–11
  fecal, 29
  and fermentation, 88–89
  genes shared by, 176
  genetic engineering of, 106
  in microbial world, 9–11
  naming of, 97–98
  niches or "professions" of, 189
  species of, 5–6, 14
  spores formed by, 166–67
  targeting for elimination, 79
Bacteria-Boosting Granola, 233–34
Bacteroides, 26, 46–47, 153–55
bananas, 184
B cells, 64, 76
Bifidobacteria, 38, 46, 49–50, 97, 98, 104–5
Big "MAC" Quesadillas, 241–42
bioreactor, 173, 177–78
birth:
  delivery method, 38, 57–58, 210–11
  microbes at, 3, 4, 10, 35
  and mother's microbiota, 37, 44
  premature, 39–42
  sterility up to, 35
Blaser, Martin, 78–80
brain, development of, 158–61
brain-gut axis, 3, 137–39, 149–52, 157, 158–61
bran, 118, 119, 131
breakfast recipes, 232–39
  Aztec Hot Chocolate, 238–39
  Bacteria-Boosting Granola, 233–34
  Morning Microbiota Smoothie, 232–33
  Muesli for Microbes, 234–35
  Symbiotic Scramble, 237–38
  Tarahumara Pacakes, 235–37
breast milk, 45–49, 50, 59, 211–12
Brownies for Your Bacteria, 262–63
Burkina Faso Skillet Cake, 263–64
Burkitt, Denis, 118–19
buttermilk, 97

cake: Burkina Faso Skillet Cake, 263–64
calories, 2, 114, 115–16
cancer, 2, 72, 76, 79, 105, 118, 128, 196–99, 218
carbohydrates, 120–23
  complex, 102, 112, 122
  GOS, 47
  hidden, 131
  HMOs, 46–48, 50, 58, 59
  in intestinal mucus, 73, 125, 217–18
  MACs, 112, 121, 122–24, 217, 218
  simple, 121–22
  total, 123–24
  what they are, 120
carnitine, 147
Cashews for Your Commensals, 249–50
castile soap, 215
CDAD (*C. difficile* associated disease), 166–67, 178–80
cell division, 21, 114
cellulose, 121
central nervous system, 3, 138–39, 160
cheese, 107
chemotherapy, 72, 196–99
Chickpea Greek Salad, 243
children:
  antibiotics fed to, 56, 57, 58, 69, 212–13
  autism of, 152–56
  brain-gut axis of, 159–60
  and diarrhea, 164
  eating habits of, 59, 221
  on farms, 67, 214
  and pets, 82, 214–15
  probiotics for, 92, 93–95, 221
  school lunches, 222–23
cholera, 22, 23–24
cholesterol, 128, 216
ciprofloxacin (Cipro), 171
cisplatin, 197–98
Cleave, Thomas, 118, 119

Clostridium difficile:
antibiotic use as risk factor for, 19, 165–67, 174, 190, 212
CDAD, 166–67, 178–80
and colonization resistance, 165
and FMT, 168–69, 178–82, 184
spores of, 166–67
colic, 49–50, 58
colon (large intestine), 2, 4–5, 13, 14
colon cancer, 32
colonization resistance, 165, 175–76
colonoscopy, 29
commensalism, 20
common cold, 96
community living, 191, 192
community supported agriculture programs (CSAs), 214
complex carbohydrates, 102, 112, 122
constipation, 118, 126, 177–78, 190
Cookies, Oatmeal, 260–61
Crohn's disease, 32, 61, 71, 105
crypts, 189
C-section births, 37–39, 41, 49, 53, 58, 211
cyclophosphamide, 196–98
cytokines, 64

Dal, Indian, 258–59
Dates, Fermented-Filling, 250–51
defensins, 92
depression, 151, 155, 160
dermatitis, 61
desserts, 260–65
Brownies for Your Bacteria, 262–63
Burkina Faso Skillet Cake, 263–64
Microbe-Friendly Oatmeal Cookies, 260–61
Middle Eastern Oatmeal Pudding, 264–65
Dethlefsen, Les, 171, 172
diabetes, 5–6, 118, 122, 128, 218
diarrhea:
and cholera, 24
gastroenteritis, 164
infectious, 164
and internal bioreactor, 177–78
pathogen-induced, from antibiotic use, 59, 96, 108–9, 165–67, 175
probiotic supplements isolated from, 104
diet, 216–25
and aging, 190–92, 204–5
gut effects of, 113, 148–49
and health, 184–85, 195, 204
and infants, 211
Japanese, 221–22
long-term patterns of, 216–17
Mediterranean, 221–22
and microbial diversity, 216
modern, 16, 70
plant-based, 218–19
digestive system:
bacteria in, 4–5, 13, 14, 29
food processing in, 11–12
of infants, 36
waste management in, 113–16
digitoxin, 202–3
dinner, 252–60
Fiber-Filled Flatbread Pizza, 255–57
Indian Dal, 258–59
Microbiota-Reframed Risotto, 257–58
Mutualist Mediterranean Soup, 253–54
Parsley Almond Pesto, 256
Sesame Seed–Crusted Salmon with Green Beans and Orange Miso Sauce, 254–55
dirt, 105–6, 215
disaccharides, 120, 122, 124
disease prevention, 107
diverticulitis, 118
DNA sequencing, 27, 28, 298
dysbiosis, 32, 44

Earth Microbiome Project, 38–39
E. coli, 73
eczema, 61, 215
Eggerthella lenta, 202–3
eggs: Symbiotic Scramble, 237–38
Eiseman, Ben, 168–69
ELDERMET, 190–92
enteric nervous system, 138

evolution, 10, 37, 46, 48, 66, 69
excretion time, 200–202
exercise, 204, 205–6

Faecalibacterium, 44, 105
Falkow, Stan, 3
fat, dietary, 119, 219
FDA (Food and Drug Administration):
  clinical trials of, 100
  and dietary fiber, 119, 125
  and FMT, 178–79
  and GRAS, 99
  and probiotics, 98, 99, 100–101, 219
fecal microbiota transplant (FMT), 168–69, 178–83, 184
Federal Trade Commission (FTC), 101
fermentation, 20, 86–89
  anaerobic metabolism, 115
  fiber for, 117
  and probiotic supplements, 102, 220–21
  and SCFAs, 216
Fermented-Filling Dates, 250–51
fermented foods:
  appendix, 269–70
  with bacteria, 88, 94–95, 112
  function of, 88–89
  pasteurized, 107–8
  probiotics in, 99
fiber, dietary, 53
  and aging, 191, 194
  Bacteroides, 26
  benefits of, 117–20
  daily recommendation, 222, 270
  and flatulence, 223–24
  lack of, in diet, 111, 117
  MAC diet, 112, 135–36, 205, 217, 218
  and microbiota diversity, 112
  nonfermentable, 126
  polysaccharides, 13
  in prebiotics, 103
  processing of, 2
  removal of, 113–14, 131
  use of term, 124–25
Fiber-Filled Flatbread Pizza, 255–57
fight-or-flight response, 149–50
Firmicutes, 76–77
fitness, 195–96
flatulence, 190, 223–24
flour, 130, 132
flu season, 213
FMT (fecal microbiota transplant), 168–69, 178–83, 184
food:
  digestion of, 11–12, 189–90
  hunting and gathering, 15, 16, 17–18, 111
  processed, 2, 16, 130–32, 220–21
  sensitivities to, 224
  and waste production, 77, 114–16
  see also diet
Food and Drug Administration (FDA), 98, 99, 100, 101, 119, 125, 178, 219
fructooligosaccharides (FOS), 103
fructose, 102–3, 120
fruits and vegetables, 122–23, 218
FTC (Federal Trade Commission), 101

galacto-oligosaccharid (GOS), 47
gas, 133–34, 190, 223–24
gastroenteritis, 164
generally regarded as safe (GRAS), 99
gene therapy, 163
genome, 19, 21, 22, 26–28, 114, 209–10
genus, 97–98
geophagia, 105–6
germ theory of disease, 23
Giardia, 164
glucose, 120, 122
glycemic index, 122
glycemic load, 122–23
Goodbelly, 107
Gordon, Jeffrey, 31–32, 42, 129
GOS (galacto-oligosaccharide), 47
Granola, Bacteria-Boosting, 233–34
GRAS (generally regarded as safe), 99
gut, as oxygen-free environment, 114–15
gut bacteria, *see* bacteria; microbes; microbiota
gut feeling, 137

hakarl, 95
hand sanitizers, 5, 68
hand-washing, 81, 83
Hadza people, 17–18, *17*
heart disease, 118, 122, 128–29, 134, 147, 148, 160, 218
Helicobacter pylori, 78–80
helminths, 70
hemorrhoids, 118
HIV infection, 72
HMOs (human milk oligosaccharides), 46–48, 50, 58, 59
homeostasis, 22
Honda, Kenya, 76–77
Hot Chocolate, Aztec, 238–39
household cleaners, 2, 5, 68, 112, 215
HPA (hypothalamic-pituitary-adrenal) axis, 138
human body, as tube, 11–14
Human Genome Project, 26–28
Human Microbiome Project, 27
human milk oligosaccharides (HMOs), 46–48, 50, 58, 59
Hummus: Hunger-Gatherer Tuber Snack, 247–48
hunter-gatherers, 15, 16, 17–18, 111
Hunter-Gatherer Tuber Snack, 247–48
hygiene hypothesis, 66–68, 69
hypothalamic-pituitary-adrenal (HPA) axis, 138

IBD (inflammatory bowel disease), 15, 71–72, 76, 104, 105, 128, 151, 177
IBS (irritable bowel syndrome), 104, 106, 151, 177
immune response, 66
immune system:
- and aging, 193–94
- on alert, 65–66, 94, 218
- autoimmune diseases, 2, 5–6, 61, 67, 69, 76
- and cancer, 196–98
- evolution of, 66
- functions of, 22, 78–80, 165, 175
- and gut microbes, 64–66, 75–77
- of infant, 36, 40, 46
- misdirection of, 63
- mobility of, 62–63
- mucosal, 70–75, *74*
- set point of, 2, 20, 80–83, 198
- systemic, 62

immunosenescence, 193–94, 203
Indian Dal, 258–59
Industrial Revolution, 16
infants:
- brain-gut axis of, 158
- breast milk for, 45–49, 50, 59, 211–12
- with colic, 49–50, 58
- C-section, 37–39, 41, 49, 53, 58, 211
- delivery method of, 38, 57–58, 210–11
- first microbiota of, 36–39, 57–59, 190, 210
- vaginal delivery, 58, 211
- weaning, 50–55, 59

inflammaging, 193–95, 203, 205
inflammation, 2–3, 44, 71–72, 73, 128, 150, 165, 216
inflammatory bowel disease (IBD), 15, 71–72, 76, 104, 105, 128, 151, 177
Insel, Thomas, 160–61
insulin, 121–22, 216
intestinal permeability, 153–54
intestinal wall, 91–92, 189
intestine, length of, 36
Inuit, 132–35, 224
inulin, 102–3, 104
irritable bowel syndrome (IBS), 104, 106, 151, 177

Japanese diet, 221–22
Japanese Popcorn, 248–49

Kashyap, Purna, 96–97, 177
kefir, 94–95, 104, 108, 213, 220, 221
Kefir Mango Lassi, 260
Kendall, Arthur, 25
kidney failure, 122
kidney function, 146–47
kimchee, 107, 112
Knight, Rob, 38–39

Koch, Robert, 24
Koch's Postulates, 24
kombucha, 95, 112

lactic acid, 90, 108
Lactobacillus, 37, 40–41, 49–50, 91, 97, 98
lactose, 88, 91, 115, 120
lactulose, 146
large intestine, 2, 4–5, 13, 14
lateral gene transfer, 176
L-carnitine, 147–48, 218–19
lecithin, 147
lemon juice, 215
Ley, Ruth, 42–44
lifestyle changes, 32–33
liver, 12, 128, 146, 200–202

MACs (microbiota accessible carbohydrates):
  "Big MAC" diet, 135–36
  and environment, 216–18
  fermentable, 115, 131, 194
  and fiber, 112, 124, 125–26, 135–36, 205, 217, 218
  and food choices, 114, 222
  function of, 121
  and SCFAs, 115, 116, 126, 205
  and vegetables, 122–23
  and Western diet, 127–28, 129, 130–32
Mango Kefir Lassi, 260
mannan, 126
Mazmanian, Sarkis, 63–64, 153
meat, red, 134–35, 147, 148–49, 218, 219
Mediterranean Diet, 221–22
Mediterranean Soup, 253–54
meningitis, 23
menus, 7-day, 227–29
metabolic syndrome, 32
metabolism, 20, 22
Metchnikoff, Élie, 89–90
methane, 133
miasmas, 23–25
mice:
  antibiotics fed to, 56–57
  anxiety in, 150
  and autism, 153–55
  and chemotherapy, 196–98
  and digoxin, 203
  gnotobiotic, 30–32
  humanized, 30
  inflammation in, 71
  lean vs. obese, 128–29, 206
  microbe-free, 30, 75, 140–42, 159
  microbial diversity of, 129–30
  microbial transplants, 142–44
  and pathobionts, 194–95
  research applicability to humans, 156–58, 198
Microbe-Friendly Oatmeal Cookies, 260–61
microbes:
  bad vs. good, 78–80
  body contact with, 68–70
  coexistence with humans, 19–22
  consuming food, 77
  custom design of, 182
  environmental, 69, 82–83, 112, 188–89, 190, 210
  imbalance of, 32
  as oldest forms of life, 10
  pathogens killed by, 165
  resistant, 68
  in supplement form, 99
microbial waste, 114–16
microbial world, 9–11
microbiology, 25
microbiome, 3–4, 19, 27–28
microbiota:
  adaptability of, 16, 80, 94, 127–28, 210, 216–17
  and aging, 190–92, 203–4
  at beginning of life, 57–59
  chemicals produced by, 3, 144–46, 183, 201
  composition of, 113
  in digestive system, 13, 14
  diversity of, 6, 18, 69–70, 111, 129, 173, 193, 214, 216, 218
  durability of, 188–89
  extinction, 111–12
  feeding, 230–31
  fluctuations of, 188

functionality of, 113
hereditary inheritance of, 188
malfunction of, 3–4
rebuilding of, 69
reprogramming, 164, 183–85
research on, 25–31
roles of, 1, 20
third trimester, 43–44
transplants of, 142–44, 168–69, 182, 184
unique, 3
Microbiota-Friendly 7-Day Menu, 227–29
Microbiota-Reframed Risotto, 257–58
Middle Eastern Oatmeal Pudding, 264–65
miso paste, 107
monosaccharides, 120, 122, 124
Moranella endobia, 21–22
Morning Microbiota Smoothie, 232–33
motility, 177–78
mucus, 14, *14*, 16
mucus layer, 73, 92, 125, 217–18
Muesli for Microbes, 234–35
multiple sclerosis, 61, 63, 78, 160
mutualism, 20
Mutualist Mediterranean Soup, 253–54

natural selection, 19, 114
necrotizing enterocolitis, 40, 96
nervous system:
and brain-gut axis, 3, 138–39, 142–52, 160–61
and fight-or-flight response, 149–50
neurotransmitters, 138
and fight-or-flight response, 149–50
nervous system, and brain-gut axis, 3, 138–39, 149–52, 160–61
nutriceutical, 95
Nut Butters, 239–41
nutritional labels, 123–25

Oatmeal Cookies, Microbe-Friendly, 260–61
Oatmeal Pudding, Middle Eastern, 264–65
obesity, 4, 5–6, 31, 32, 104, 128–29, 160, 211, 216, 218
obsessive-compulsive disorder, 155
oligosaccharides, 121, 212
FOS, 103
GOS, 47
HMOs, 46–48, 50, 58, 59
omnivores, 134
Orange Miso Sauce, 254–55
oxaliplatin, 197–98

Pancakes, Tarahumara, 235–37
pancreas, 12
Parsley Almond Pesto, 256
Pasteur, Louis, 23
pathobionts, 194–95, 219
pathogens, 164–65, 173–77
antagonistic interaction of, 19
as "bad" bacteria, 22–23, 24, 25
and colonization resistance, 165
defenses against, 62–63, 73, 173–74, 218
and germ theory of disease, 23
and immune system, 62–63, 66
persistence of, 175
PB&J 2.0, 239–40
p-cresol, 201–2
pectin, 121, 126, 145, 189
penicillin, 170
pesticides, 81, 214
Pesto, Parsley Almond, 256
pets, 82, 112, 214–15
phagocytes, 89
pharmaceuticals, 199–203
pickles, 95, 107, 112, 220
pizza: Fiber-Filled Flatbread Pizza, 255–57
plasticity, 16, 164
Pollan, Michael, 5
polysaccharides, 13, 102, 120, 121, 124, 126
Popcorn, Japanese, 248–49
porphyranase, 126
prebiotics:
in baby formula, 59
inulin, 102–3, 104
pregnancy, 42–44, 46, 211

Probiotic Pick-Me Up, 248–49
probiotics, 85–86, 219–21
  and aging, 205
  in baby formula, 47–48, 59, 213
  for children, 92, 93–95, 221
  erratic effects of, 94
  following antibiotic use, 108, 221
  future of, 104–6
  health claims for, 93–95, 96–97, 100–101
  immune boosting, 76, 86, 93–94
  promotion of, 95–97, 99–100, 101, 219
  and regularity, 109
  transience of, 86, 90–93
  use of term, 86, 98
  user's guide, 107–10
Proteobacteria, 38, 44, 49–50
psychobiotics, 151

Quesadillas, Big "MAC," 241–42

recipes, 232–65
  breakfast, 232–39
  desserts, 260–65
  dinner, 252–60
  school lunches, 239–42
  snacks, 246–52
  work lunch, 242–46
refrigerators, 86–88
Relman, David, 171, 172
RePOOPulate, 181–83
rifaximin, 146
Risotto, Microbiota-Reframed, 257–58
Roach, Mary, *Gulp: Adventures on the Alimentary Canal*, 11

salads:
  Chickpea Greek Salad, 243
  Soba Noodle Salad with Probiotic Peanut Miso Sauce, 243–45
salmon: Sesame Seed–Crusted Salmon with Green Beans and Orange Miso Sauce, 254–55
Salmonella, 19, 64, 164, 165, 173, 174–75, 212
Salyers, Abigail, 25–26, 46
Sanders, Mary Ellen, 96, 101, 104
sauerkraut, 95, 107, 108, 112, 220
schizophrenia, 155
school lunches, 222–23, 239–42
  Almond/Walnut Butter, 240–41
  Big "MAC" Quesadillas, 241–42
  PB&J 2.0, 239–40
Scrambled Eggs, Symbiotic, 237–38
seaweed, 126–27
serotonin, 139
Sesame Seed–Crusted Salmon with Green Beans and Orange Miso Sauce, 254–55
7-Day Menu, 227–29
Shigella, 184
short-chain fatty acids (SCFAs), 115–17, 146, 218
  and aging, 194
  and bacterial diversity, 216
  functions of, 77, 103, 115, 116–17
  and MACs, 115, 116, 126, 205
  and polysaccharides, 121
  in research, 128, 191
  and T-reg cells, 77
simple carbohydrates, 121–22
skinny fat people, 129
small intestine, 12–13, 120
Smoothie, Morning Microbiota, 232–33
snacks, 246–52
  Cashews for Your Commensals, 249–50
  Crunchy Yogurt Parfait, 252
  Fermented-Filling Dates, 250–51
  Hunter-Gatherer Tuber Snack, 247–48
  Japanese Popcorn, 248–49
  Plain Yogurt, 251–52
Soba Noodle Salad with Probiotic Peanut Miso Sauce, 243–45
social network, 204, 206–7
soil microbes, 68, 83
Soup, Mutualist Mediterranean, 253–54
sour cream, 94, 104, 107
starch, 120–21
starter cultures, 109

statins, 201
sterile environment, 213–15
stool, ideal, 110
stool banks, 179, 181
stool samples, 29
Strachan, David, 66
stress, 150–51
stroke, 122, 147, 218
succession, 45
sucrose, 120
superbugs, 81, 170, 176–77, 213
swab inoculation, 39
sweeteners, 221
symbiosis, 20, 22
Symbiotic Scramble, 237–38
synbiotics, 103–4
synergism, 103

Tabbouleh, Totally MAC-Filled, 245–46
Tarahumara Pancakes, 235–37
T cells, 62–63, 64, 76
thrush (yeast infection), 41
TMAO (trimethylamine-*N*-oxide), 134–35, 147–49, 218–19, 222
Totally MAC-Filled Tabbouleh, 245–46
Toxoplasma gondii, 140
transfaunation, 169
T-reg, 76, 77
Tremblaya princeps, 21–22
Treponema pallidum, 140
trimethylamine-*N*-oxide (TMAO), 134–35, 147–49, 218–19, 222
Trowell, Hugh, 119
tuberculosis, 23, 24
twins, fraternal, 45
Tylenol (acetaminophen), 200, 201–2

ulcerative colitis, 61, 71, 105, 125
ulcers, 79
USDA dietary guidelines, 222

vaginal delivery, 58, 211
vaginal swab, 58
van Gogh, Vincent, 202
vegetables and fruits, 122–23, 218
vegetarians; vegans, 134, 147–48, 218
Vibrio cholera, 19, 24
villi, 12, *12*
vinegar, 215
Vos, Willem de, 49

Walker, Alec, 118–19
weaning, 50–55, 59
Western lifestyle, 15–18, 112, 213–15, 225–26
wheat berry, 130
work lunch, 242–46
  Chickpea Greek Salad, 243
  Soba Noodle Salad with Probiotic Peanut Miso Sauce, 243–45
  Totally MAC-Filled Tabbouleh, 245–46

yeast, 115
yeast infection (thrush), 41
yogurt, 219–20
  and gut bacteria–brain connection, 157
  probiotics in, 94, 98, 99, 213
  recommended as fermented food, 88, 99, 107, 108, 112, 115, 219
  research with, 156–57
  synbiotics promoted in, 104
Yogurt, Plain (recipe), 249
Yogurt Parfait, Crunchy, 250